ENERGY CONSCIOUS DESIGN

A primer for architects

EDITED BY

JOHN R.GOULDING J. OWEN LEWIS THEO C.STEEMERS

Commission of the European Communities

This publication has been prepared in the Third Solar R+D Programme of the Commission of the European Communities Directorate-General XII for Science, Research and Development, within the SOLINFO Action coordinated by the Energy Research Group, University College Dublin. Original text and diagrams were prepared by the team Architecture et Climat, Centre de Recherches en Architecture, Université Catholique de Louvain, Belgium, under the direction of Professor André De Herde.

Publication arrangements have been made under the VALUE Programme (specific programme for the Dissemination and Utilization of Community research results) within the Research Dissemination: Energy Efficient Building project of the Commission of the European Communities DGXIII.

The Louvain team members who contributed to this book are:

B. Adriaens
M . Boisdenghien
E. Gratia
B. Meersseman
Mrs M.H. Dehut, Secretary
M. Saelen, Computer Scientist

Editorial Advisory Group:

Patrick Achard, Valbonne
Alex Lohr, Köln
Albert Mitja i Sarvise, Barcelona
Martin de Wit, Eindhoven

Additional Contributions from:

Professor John Page, Sheffield
Marc Fontoynont, Lyon

Revision of text by:
Judith Stammers, London

Design and principal photography by W.H. Hastings, Dublin

Graphical illustration and pagemaking by:
W.H. Hastings, John Kelly and Pierre Jolivet, Dublin

Additional assistance by:
Mary Rigby and Kay Dunican

ENERGY CONSCIOUS DESIGN

A PRIMER FOR ARCHITECTS

EDITED BY

JOHN R.GOULDING J. OWEN LEWIS THEO C.STEEMERS

Written by

Architecture et Climat,
Centre de Recherches en Architecture,
Université Catholique de Louvain, Belgium.

Produced & coordinated by

The Energy Research Group,
School of Architecture,
University College Dublin.
Richview, Clonskeagh,
IRL-Dublin 14.

Publication No. EUR 13445 of the Commission of the European Communities, Scientific and Technical Communication Unit, Directorate-General Telecommunication, Information and Innovation, Luxembourg.

Typeset by
the Energy Research Group, School of Architecture, University College Dublin.
and printed in Singapore

Published by
B.T. Batsford Ltd, 4 Fitzhardinge Street, London W1H 0AH

for the
COMMISSION OF THE EUROPEAN COMMUNITIES
Directorate General XII for Science,
Research and Development
within the SOLINFO Action of the
Solar Energy Applications to Buildings Programme
managed by Theo C. Steemers at the Commission
and
Directorate General XIII for Telecommunications,
Information Technology and Innovation
within the Research Dissemination: Energy Efficient Building project of the VALUE programme (specific programme for the Dissemination and Utilisation of Community research results).

A catalogue record for this book is available from the British Library.

ISBN 0 7134 69196

CONTENTS

PREFACE

This handbook represents an important strand in the Commission of the European Community's programme to promote the design and construction of energy-efficient, passive solar buildings in Europe.

It has evolved from an earlier 'Preliminary' edition printed in a limited run in 1986 which has provided a significant amount of material for this edition. However, the enlargement of the Community with the accession of Greece, Portugal and Spain together with progress in research has prompted a revised edition of the handbook with a new balance of emphasis on building design for northern and southern European climates combined with extensive rewriting and increased coverage.

In addition to a new section on natural cooling, certain other topics including daylighting and thermal comfort are covered in greater depth.

The format of this revised edition has been redesigned, separating illustrated introductory texts from design information and providing two distinct but logically consistent volumes, 'Energy Conscious Design - A primer for European architects' and 'Energy in Architecture - The European Passive Solar Handbook'. This approach has allowed the content of the handbook to be tailored more appropriately in its complexity and presentation to the different needs of building designers who wish to inform themselves about passive solar design and those engaged in the task of designing or constructing passive solar building. It is anticipated that both volumes will be translated into other EC languages.

Energy Conscious Design - A Primer for Architects is in the form of an illustrated textbook, covering the principles of climatic design, passive solar heating, natural cooling, the optimal use of available daylight and an overview of the factors to be considered in providing conditions of thermal and visual comfort for building occupants.

For those unfamiliar with the subject it is strongly recommended as an introduction to the more detailed, design-oriented material in its companion publication, 'Energy in Architecture: The European Passive Solar Handbook'.

Dr. W. PALZ
Division Head "Renewable Energies"
Commission of the European Communities

INTRODUCTION

To make a building is to create a system linked to its surrounding environment, and subject to a range of interactions affected by seasonal and daily changes in climate and by the varying requirements of occupants in time and in space.

Some twentieth century buildings seek to deny these inevitable interactions and subdue them with expensive heating, cooling and lighting equipment. A more climate-sensitive approach is proposed here which recognises and responds to seasonal and daily changes in the environment for the well-being and comfort of the occupants. The relationship between people, their living place and the environment is re-examined and resolved in an architecture which permits a dynamic interaction to occur.

In recent years these issues have most frequently been addressed in the design process after the building form has been fixed, and we have become used to thinking of heating, cooling and lighting devices as add-on equipment to be sized and placed in more or less completed buildings. While this may be a pragmatic or convenient approach, it diminishes the opportunity to design, at a more holistic level, buildings which can respond to the environment by virtue of their form and the intelligent use of materials with minimal reliance on machinery. Rediscovery of this design skill adds a dimension to the design process which offers sound parameters as generators of architectural form.

To achieve this calls for a knowledge of climate and an awareness of the available technologies which can be employed in building, combined with an understanding of what constitutes comfort and discomfort and how these conditions can be affected by changes in climate. These issues are relevant to all buildings and locations whether the predominant need is for heating, cooling or daylighting. Initial design decisions will focus on the location of the building, its basic form, the arrangement of the spaces, the type of construction and the quality of the environment to be provided, resulting in an architectural response of high quality which is in harmony with its environment.

In most situations it is necessary to provide some additional heating or cooling at certain times. Similarly, daylighting cannot meet all lighting requirements and therefore these auxiliary inputs and their control must be addressed once the contributions by natural means and the patterns of use are known.

The design and construction of a building which takes optimal advantage of its environment need not impose any significant extra cost, and compared to more highly-serviced buildings it may be significantly cheaper to run.

There are two major strategies, depending on the regional climate and the predominant need for heating or cooling:

- in cold weather - maximize 'free' heat gains, create good heat distribution and suitable storage within the building and reduce heat losses while allowing for sufficient ventilation;
- in warm weather - minimize heat gains, avoid overheating and optimize cool air ventilation and other forms of natural cooling.

To the above must be added a daylighting strategy. The availability of daylight is influenced by latitude and climate. The use of natural light to replace electrical light is particularly important in large buildings with a low surface-to-volume ratio, where unwanted internal heat gains caused by artificial lighting can be considerable and often may require the use of mechanical air-conditioning.

Information on climate may be considered on three levels: macroclimate, mesoclimate, and microclimate. Macroclimatic data are gathered at meteorological stations and describe the general climate of a region, giving details of sunshine, wind, humidity, precipitation and temperature etc. Mesoclimatic data, although sometimes more difficult to obtain, relate to the modification of the macroclimate or general climate by established topographical characteristics of the locality such as valleys, mountains or large bodies of water and the nature of large-scale vegetation, other ground cover, or by the occurrence of seasonal cold or warm winds. At the microclimate level we can consider the human effect on the environment and how this modifies conditions close to buildings. For example, planted vegetation and neighbouring buildings influence a site's exposure to the sun and wind. Water and vegetation affect humidity and city planning modifies wind direction, intensity and air temperature.

All buildings have a primary function to enclose space to provide an internal environment suitable for habitation. This in turn provides an opportunity to create sheltered, comfortable spaces around the building which can have a significant amenity value at certain times of the year, and these external spaces can have a beneficial effect on the internal environment of the building by minimizing the need for artificial heating or cooling. The configuration of the building and the arrangement of spaces according to function becomes important as does the selection of systems or devices to control the natural heating or cooling of the building.

In essence then, in cold periods the need is to collect and store heat energy for distribution when and where there is a need; for example, at night or to north facing rooms. Insulation is also required to allow heat to be retained.

In warm periods the need is to avoid overheating from direct solar radiation or unwanted internal heat gains from appliances or people, and to dissipate heat by natural ventilation or other means of natural cooling.

The use of south facing glazing, particularly in the form of a conservatory or attached sunspace as it is sometimes called, can in addition to the amenity value it provides, act as both a collector of heat and as a 'buffer zone' which can insulate part of the building when there is no heat from the sun. This dual role can be compared with the 'north facing buffer zone' which acts solely as a protection against heat loss.

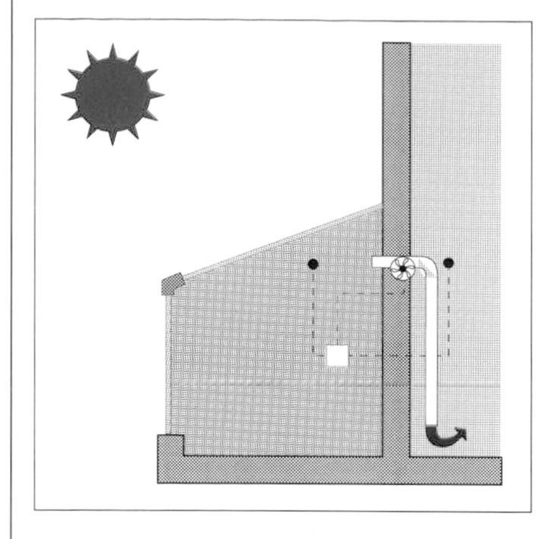

Once captured, heat may be stored within the structure of the building making use of the thermal inertia of heavy mass elements such as walls or floors without causing any overheating, and may be distributed throughout the building when needed. The design of insulation, which may be movable, can play an important role in retaining heat and controlling its storage and distribution.

Fixed or movable devices may be used to prevent overheating from direct solar radiation. Again, the thermal inertia of the building structure may be used to reduce overheating and regulate internal temperatures. Unwanted heat from appliances or occupants may need to be controlled and heat dissipated by ventilation or other forms of natural cooling.

Good daylighting design will optimize the collection of natural light, ensuring its distribution about the building to provide lighting levels appropriate to each activity while avoiding visual discomfort associated with high contrast or glare.

The contribution which daylight can make to energy saving, visual comfort and the quality of the thermal living or working environment is relevant in all climates, whether the predominant need is for heating or cooling.

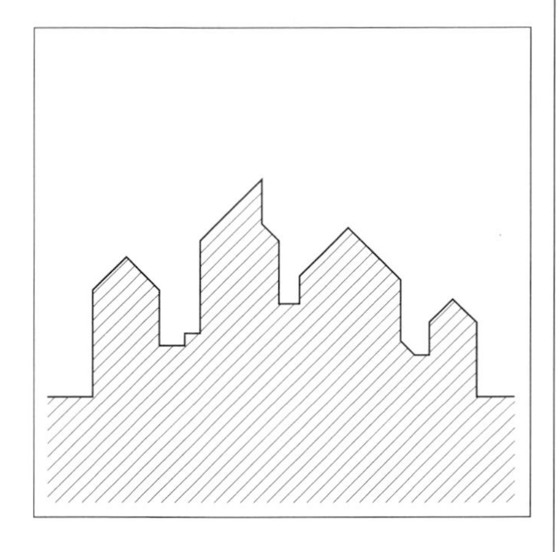

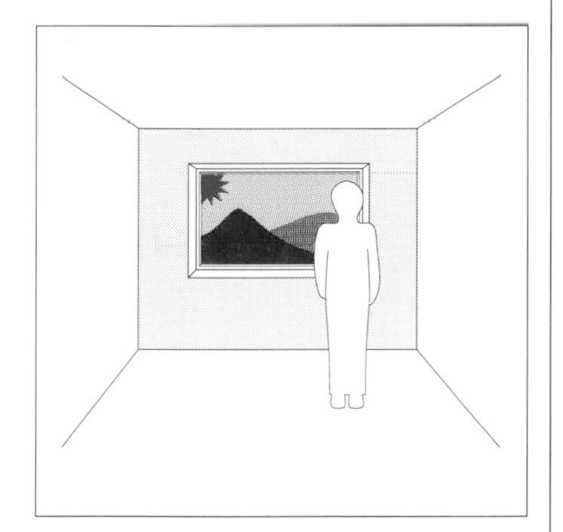

The design of an isolated building and its immediate environment, where it is unaffected by neighbouring buildings, is one matter. However, more often, one must consider the negative and positive aspects of building in urban locations. Clusters of buildings create their own microclimates through shading, shelter, wind deflection, and the emission of heat. Depending on the climate, some of these aspects may be turned to advantage while others may have to be minimized. The interactions between buildings are by their nature very complex and while some design tools and guidelines exist to help the designer to understand the phenomena involved and predict how the building will perform, the development of more elaborate, often computer-based tools continues.

In order to provide conditions of thermal comfort a knowledge of the range of people's comfort tolerances is needed. Thermal comfort is affected by such factors as temperature, humidity, airflow (draughts), the level of physical activity, the amount of clothing being worn and even the weight of the individual concerned. Comfort is to some extent subjective and consequently a capacity for the individual to have some control over his or her environment is desirable. The section on Comfort characterises the range of comfort conditions while the section on Behaviour gives pointers to the way people respond in buildings and how occupant behaviour may affect the building's performance.

This book attempts to demonstrate the benefits of an approach to the design of buildings and their immediate surroundings which takes advantage of natural phenomena instead of fighting the influences of nature with expensive and often environmentally-destructive heating, cooling or lighting equipment and the energy they consume. The overall goals to which the book is directed are improved thermal and visual comfort in more environmentally-benign buildings, and the synthesis of these objectives in good architectural design. What is proposed is a fundamentally more thorough approach to building design which, while adopting additional performance parameters, offers the possibility of exciting new architectural design opportunities.

PART 1

THE CLIMATE

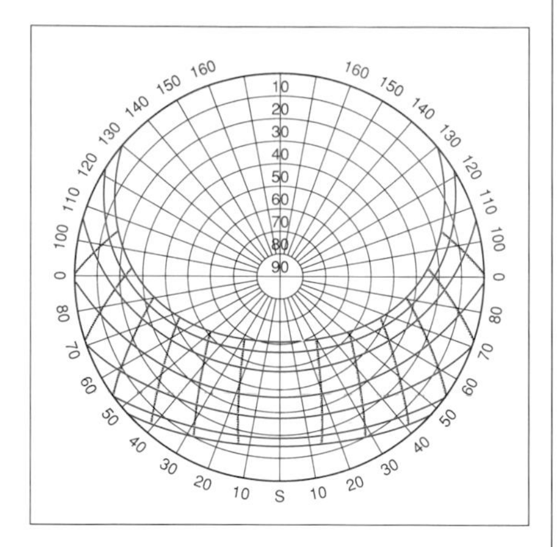

INTRODUCTION

From the viewpoint of human comfort and energy use, the climatic condition of a place can be divided broadly into negative and positive effects. In general, the aim of climate-conscious architecture is to provide protection from the negative factors and take advantage of the positive ones in order to meet the comfort requirements of the inhabitants and secure an economical level of energy consumption. A year-round analysis is required. Buildings have to perform appropriately in summer, winter and intermediate seasons. The latter are often characterized by great day-to-day variability, demanding flexibility of operation. To allow the architect to analyse the climate at a particular site, the climatic factors need to be expressed quantitatively. The key factors are the position of the sun, the amount of solar energy, the air temperature, long wave radiation and wind condition. Humidity, best assessed as the water-vapour pressure, is also an important factor in hot weather.

The geometric range of the sun's position throughout any particular day determines the directly-radiated surfaces. The range changes from season to season. The strength of the direct solar beam on a specific surface is described by its beam irradiance. This is the heat flow per unit area and is expressed in watts per square metre (W/m^2). In Europe at low level sun, the beam irradiance seldom exceeds 900 W/m^2. As well as the direct solar beam, building surfaces also receive diffused solar radiation scattered by the air and particles in the sky and reflected from clouds, ground, adjacent buildings, etc. If the solar energy reaching a surface is summed over a stated period of time, the irradiation is obtained.

The outdoor air temperature describes the thermal ambience of a place. It can be measured using thermometers shaded by a well-ventilated whitened meteorological screen mounted at a standard height. Air temperature is a major factor in determining thermal comfort.

Knowledge of wind conditions makes it possible to assess the impact wind can have on ventilation and heat loss. Information on typical wind speeds and directions is needed to devise protection against adverse winds in winter. In summer, positive use of wind for natural cooling is important, especially in southern Europe. Wind speeds are very dependent on height above ground. Substantial deflections are caused by buildings.

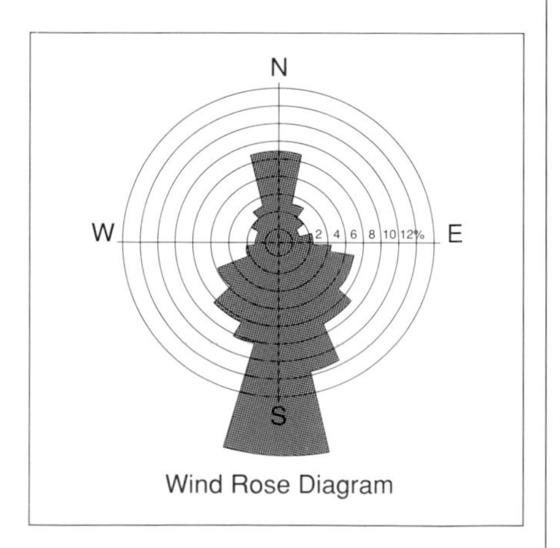

Wind Rose Diagram

The humidity of the air exerts a strong influence on comfort in hot weather. High water vapour pressure makes it difficult for the body to evaporate sweat at adequate rates to stay cool.

The available observed values of the above climatic factors are those measured at meteorological stations where the instruments are mounted under standardized conditions at standard heights above level, grass-covered ground. These measured values provide the macroclimate database for the area. Sometimes, access to more local data can help the architect to understand the conditions of the site more fully.

The architect must always interpret the required climate data according to the place where he wants to build. He or she must consider the mesoclimate – the modifications which topography and vegetation make to the macroclimate of the region. He or she must also take into account the microclimate – the effect on the mesoclimate of what man does to the local environment.

Wind flow is strongly affected by topography – for instance, by the shelter provided by hedgerows and trees, by the surface properties of features such as lakes or shrubs, and by the shape of the ground. Wind flow is also affected by nearby buildings: in towns, it can be highly perturbed. The temperature microclimate is influenced by the wind microclimate. It is also affected by the nature of the ground cover. For instance, dark pavements heat up in the sun whereas green grass produces a much cooler environment. The solar radiation climate at the site is affected by the presence of pollution.

This chapter begins by explaining the main climatic features found in Europe's major climate zones. Information is then given about the changes which have to be applied to the general values to provide mesoclimate data for the local area. Finally, the last section of the chapter examines the microclimate experienced at sites due to man-made modifications to the environment. Microclimate data can be estimated but it is desirable to complement this with direct measurement on site. It can be helpful for the architect to note the way in which the vernacular architecture and natural features handle or are affected by the local climate conditions. It can also be useful to collect information about the site from local inhabitants – although care should be taken that the latter do not exaggerate the frequency of extreme conditions !

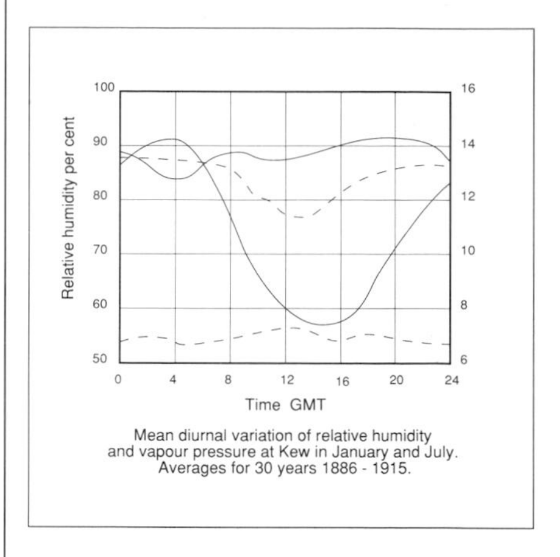

Mean diurnal variation of relative humidity and vapour pressure at Kew in January and July. Averages for 30 years 1886 - 1915.

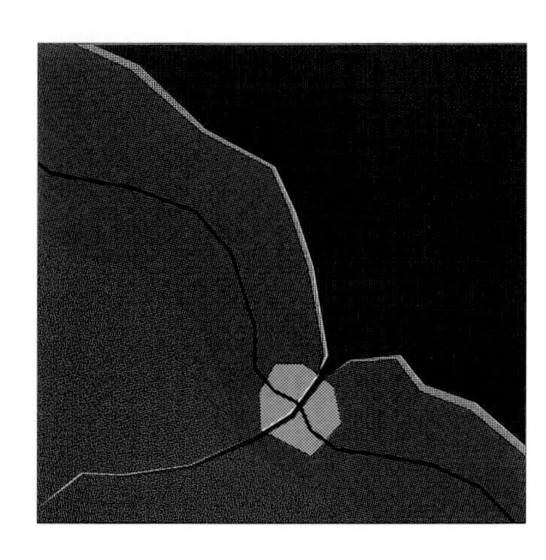

CHAPTER 1

THE CLIMATE
MACROCLIMATE

Macroclimate data, which is provided by standard meteorological stations, describes the general character of a region in terms of sunshine, cloud, temperature, wind, humidity and precipitation. An understanding of the climate is essential to the design of climate-responsive buildings.

CLIMATE CONDITIONS IN EUROPE

Most of the European Community is located in a climatic region which is more or less temperate and is bounded by areas with strongly contrasting physical characteristics – the Atlantic Ocean to the west, the Arctic Sea to the north, a large continental land mass to the east and the Mediterranean Sea and North Africa deserts to the south.

The winds coming from the west are usually relatively humid. When westerly flows from the Gulf Stream are present in winter, the warm humid air causes relatively mild, humid and cloudy weather. In summer, the westerly flows remain humid but the air is usually cool and fresh compared with the east winds. The westerly flows normally carry much less dust so that when the sun shines through gaps in the commonly-occurring clouds the beam strength is rèlatively high. The diffuse radiation caused by reflection in clouds is often strong under broken cloud conditions.

The north winds bring cold, dry air from the Arctic.

In winter, the east winds (generated under the direct influence of the Siberian anticyclones) bring about the cold winters of Central Europe. The air then is cold but dry with fewer clouds (and therefore more solar radiation) than with westerly flows. In summer, the easterly continental air flows tend to be warm and dusty, reducing the strength of the solar beam.

Winds in the southern and eastern areas of the Community are strongly influenced by the blocking effects of the Alps.

While more detailed information can be obtained from maps which deal specifically with wind, solar radiation, temperature, humidity and precipitation, a combination of these provides a broad characterization of the climate in Europe:

1. **Northern European Coastal zone:**
   Cold winters with low solar radiation and short days, mild summers.
2. **Mid European Coastal zone:**
   Cool winters with low solar radiation, mild summers.
3. **Continental zone:**
   Cold winters with high solar radiation and longer days, hot summers
4. **Southern and Mediterranean zone:**
   Mild winters with high solar radiation and long days, hot summers.

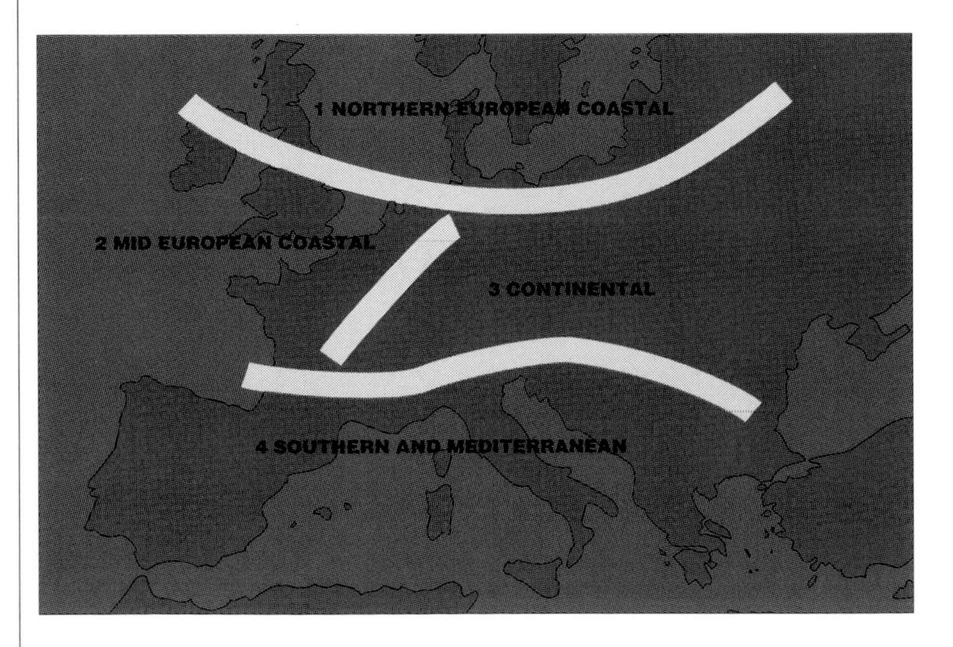

THE SUN: POSITION

The position of the sun in the sky, and hence the direction of the solar beam, is described by the solar altitude and solar azimuth angles.
The solar altitude (γ) is the angle between the line to the centre of the sun and the horizontal plane. When the sun is on the horizon, the solar altitude is 0 degrees. When the sun is directly overhead, it is 90 degrees.
The azimuth (α) is the angle between true south and the point on the horizon directly below the sun. By convention, it is negative before noon and positive after noon.

The altitude and azimuth angles vary from hour to hour and season to season.
At the northern latitude summer solstice (21 June), the sun's rays make an angle of 23°27' to the equatorial plane. The beam is approximately perpendicular to the Tropic of Cancer (22°30'N). In the northern hemisphere, the day length reaches its maximum value then. North of latitude 23°27'N, the noon solar altitude is at its greatest value for the year.
At the winter solstice (21 December), the sun's rays make an angle of -23°27' to the equatorial plane. The beam radiation is approximately perpendicular to the Tropic of Capricorn (20°30'S). In the northern hemisphere, the days are at their shortest and the noon solar altitudes have their lowest values.
At the northern hemisphere spring and autumn equinoxes (21 March and 22 September), the sun's rays are perpendicular to the equator. The day and night lengths are almost equal everywhere in the world.

The solar altitude and azimuth for the whole year, hour-by-hour, can be plotted on a solar chart. The altitude scale is shown on a series of concentric circles. The azimuth scale is set around the perimeter of the chart. The azimuth angle is read by setting a straight edge from the centre of the chart to the intersection of the required hour and date path lines and noting where it cuts the chart perimeter. Different charts are required for different latitudes.

In Europe, the extreme values of the sun's position are as follows:
- in the south of Europe (for example, in southern Greece) at latitude 36 N, the sun's path is 240 degrees wide at the summer solstice and the maximum solar altitude is 77 degrees. At the winter solstice, the sun's path is 120 degrees wide and the maximum solar altitude is 30 degrees.
- in the north of Europe (for example, in Denmark or Scotland) at latitude 56 N, the sun's path is 270 degrees wide at the summer solstice and the maximum solar altitude is 58 degrees. At the winter solstice, the sun's path is 90 degrees wide and the maximum solar altitude is 11 degrees.

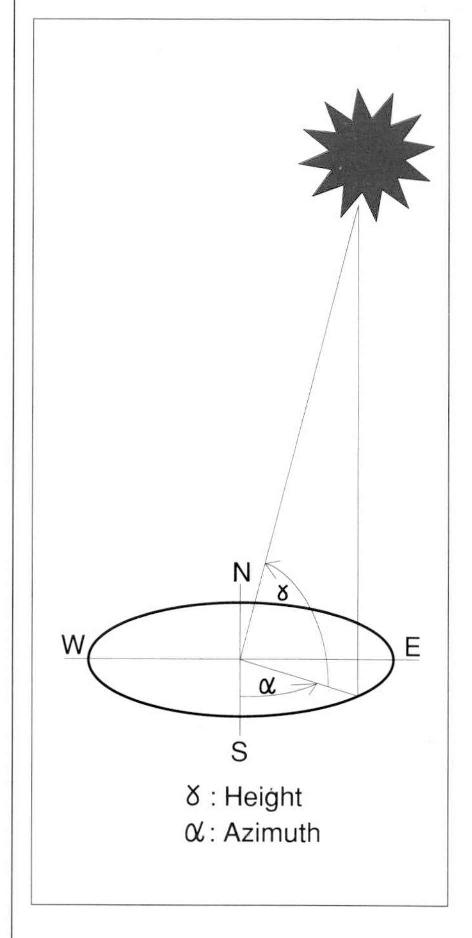

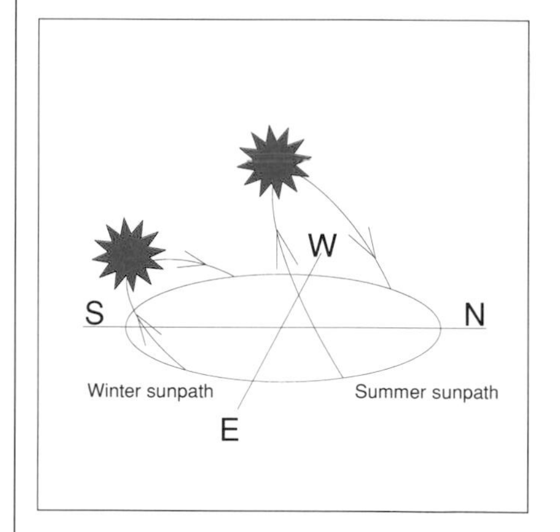

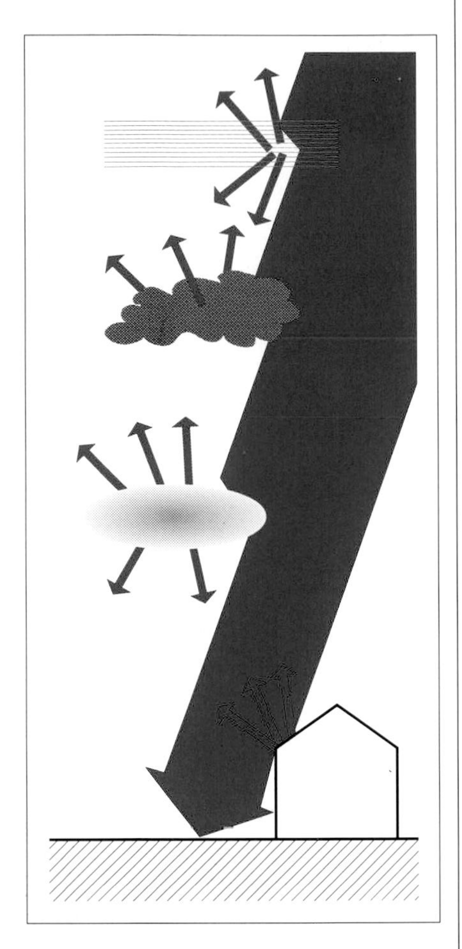

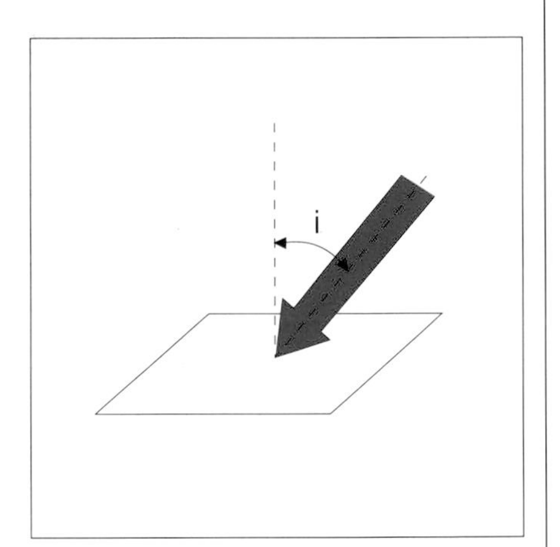

THE SUN: RADIATION

The amount of solar radiation reaching the ground depends on the composition of the atmosphere and the path length of the beam. As the beam radiation passes through the atmosphere it is partly scattered by air molecules, dust particles and water droplets and partly absorbed by water vapour, ozone, carbon dioxide and other gases. Clouds, in particular, cause absorption and scattering. Over 60% of the radiation reaching horizontal surfaces at high latitudes in the European Community is diffuse, as a result of such scattering. The proportion of diffuse radiation is smaller in southern Europe. The longer the path length through the atmosphere and the greater the amount of water vapour and dust particles, the weaker the solar beam.

The sum of the direct (or beam) and diffuse solar irradiation on a horizontal surface constitutes the globally available energy. The quantity of available energy due to solar radiation, sometimes called the energetic exposure, is a function of the irradiance.

The solar irradiance is the amount of radiant energy from the sun falling on a square metre of surface at any instant. It is usually measured in Watts per square metre and, as indicated above, has two components, the beam component and the diffuse component.

- The beam irradiance falling on a given surface (G_b) depends on the angle of incidence between the sun's rays and the normal (a line at 90°) to the surface.
- The diffuse irradiance (G_d) is the sum of the diffuse irradiance received from the sky after being reflected from the clouds. To this can be added the diffuse irradiance reflected from the ground, the neighbouring landscape and adjacent buildings.

The sum of the direct (beam) and diffuse irradiation on the surface is known as the global irradiance (G).

Irradiance varies from moment to moment. It is dependent on geographical area, latitude, season, time of day and meteorological conditions. If the irradiance on a surface is integrated over a stated period of time, the irradiation is obtained. The commonly-used period of integration is the day, so typically irradiation is given in Kilowatt hours (kWh) per square metre per day.

In Europe, the value for the annual mean daily irradiation on a horizontal surface varies from 2.25 kWh per square metre in Scotland to 6 kWh per square metre per day in the Mediterranean area. Data for horizontal surfaces at meteorological stations throughout the Community can be found in Volume I of the Commission of the European Communities' ***European Solar Radiation Atlas***.

Inclined surfaces receive different amounts of daily irradiation to horizontal surfaces. Tilting a surface towards the mean position of the sun increases the daily irradiation. In addition, the colour of the ground influences the daily slope value because it affects the amount of radiation reflected from the ground onto the inclined surface. Maps of calculated irradiation data for inclined surfaces throughout the European Community are given in Volume II of the ***European Solar Radiation Atlas***. In preparing the maps a ground albedo (the proportion of the incident solar radiation reflected from the ground) of 0.2 was assumed.

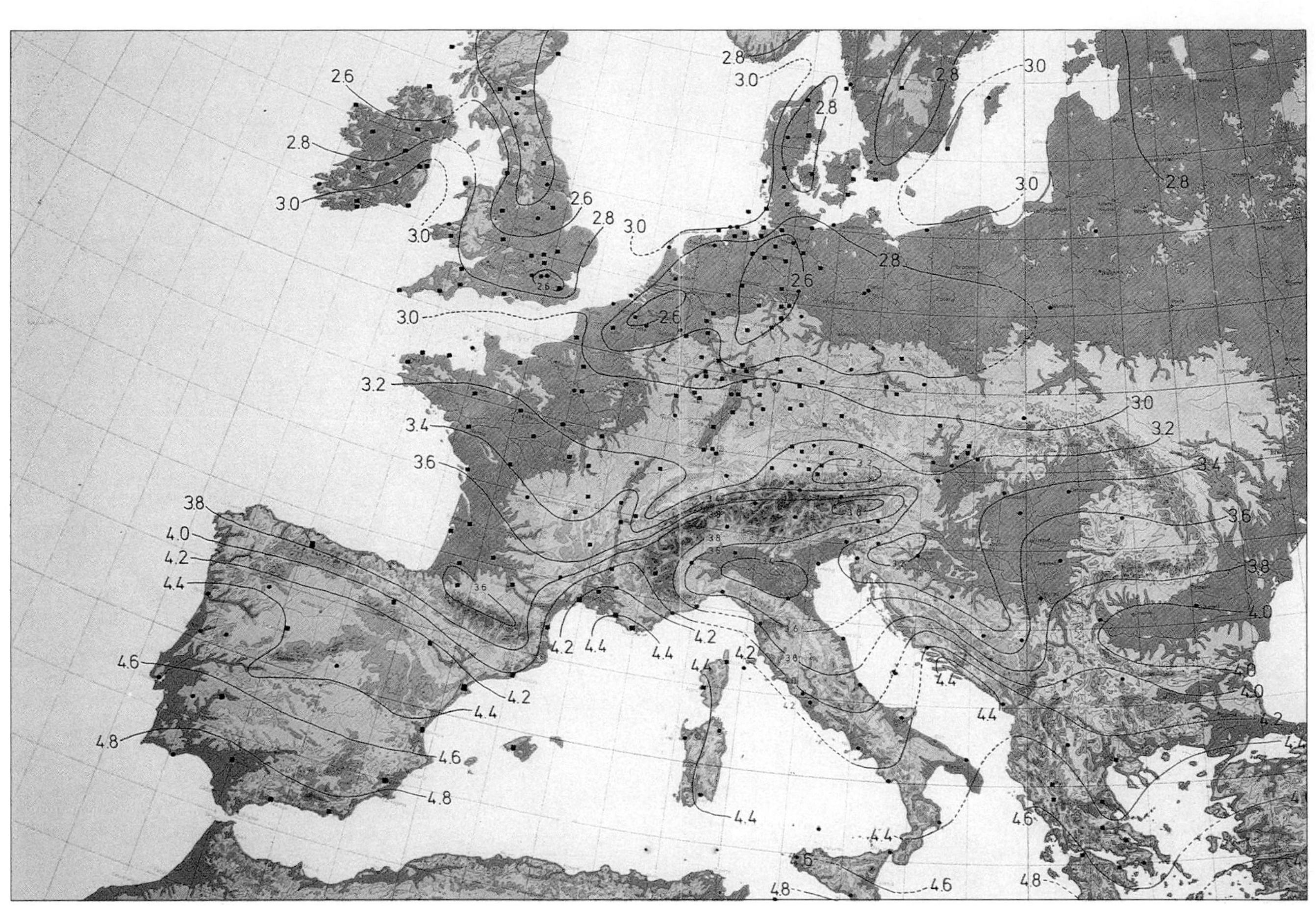

ANNUAL Daily global irradiation in kWh/m² : Mean annual means 1966 - 1975

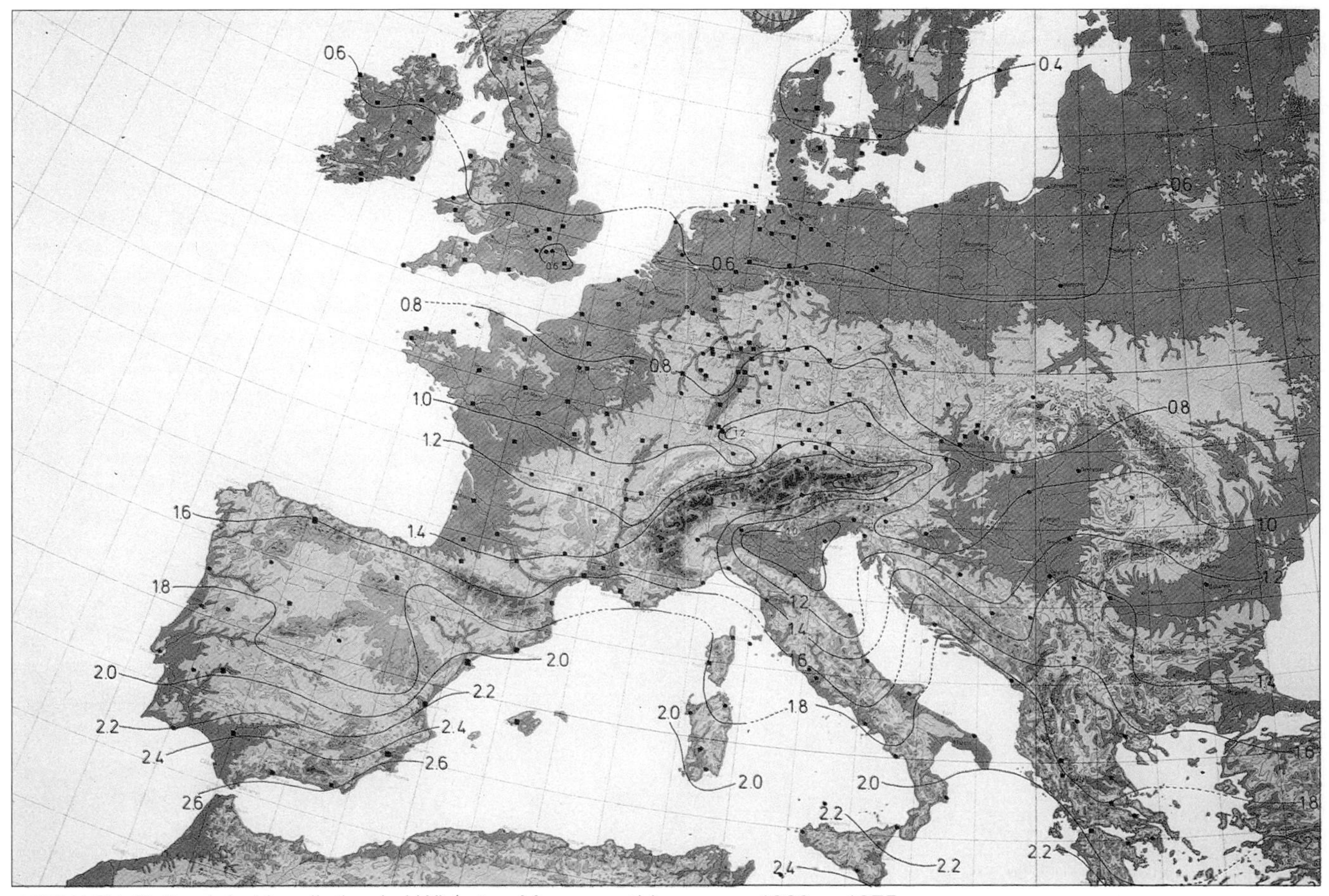

JANUARY Daily global irradiation in kWh/m² : Mean monthly means 1966 - 1975

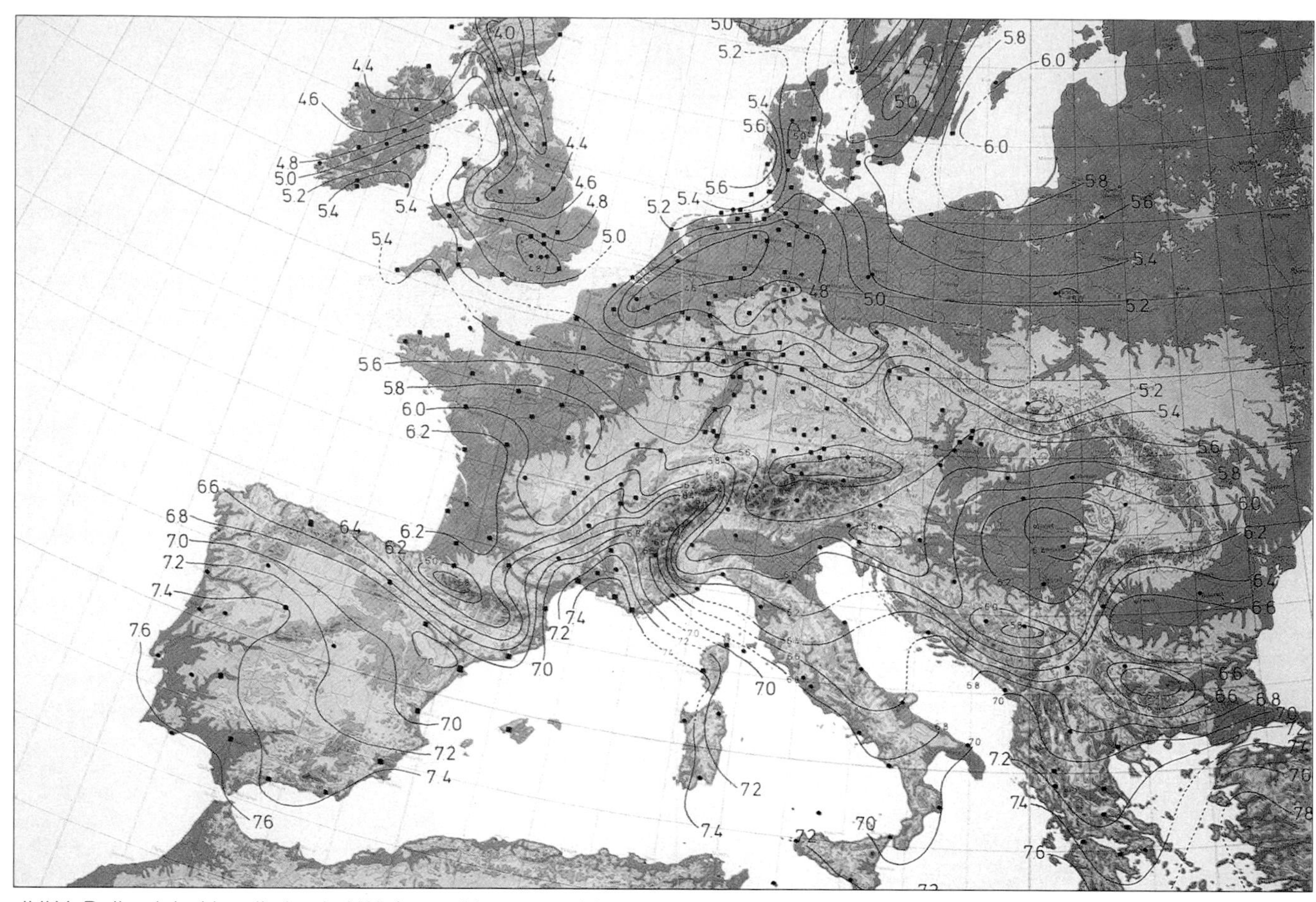

JULY Daily global irradiation in kWh/m2 : Mean monthly means 1966 - 1975

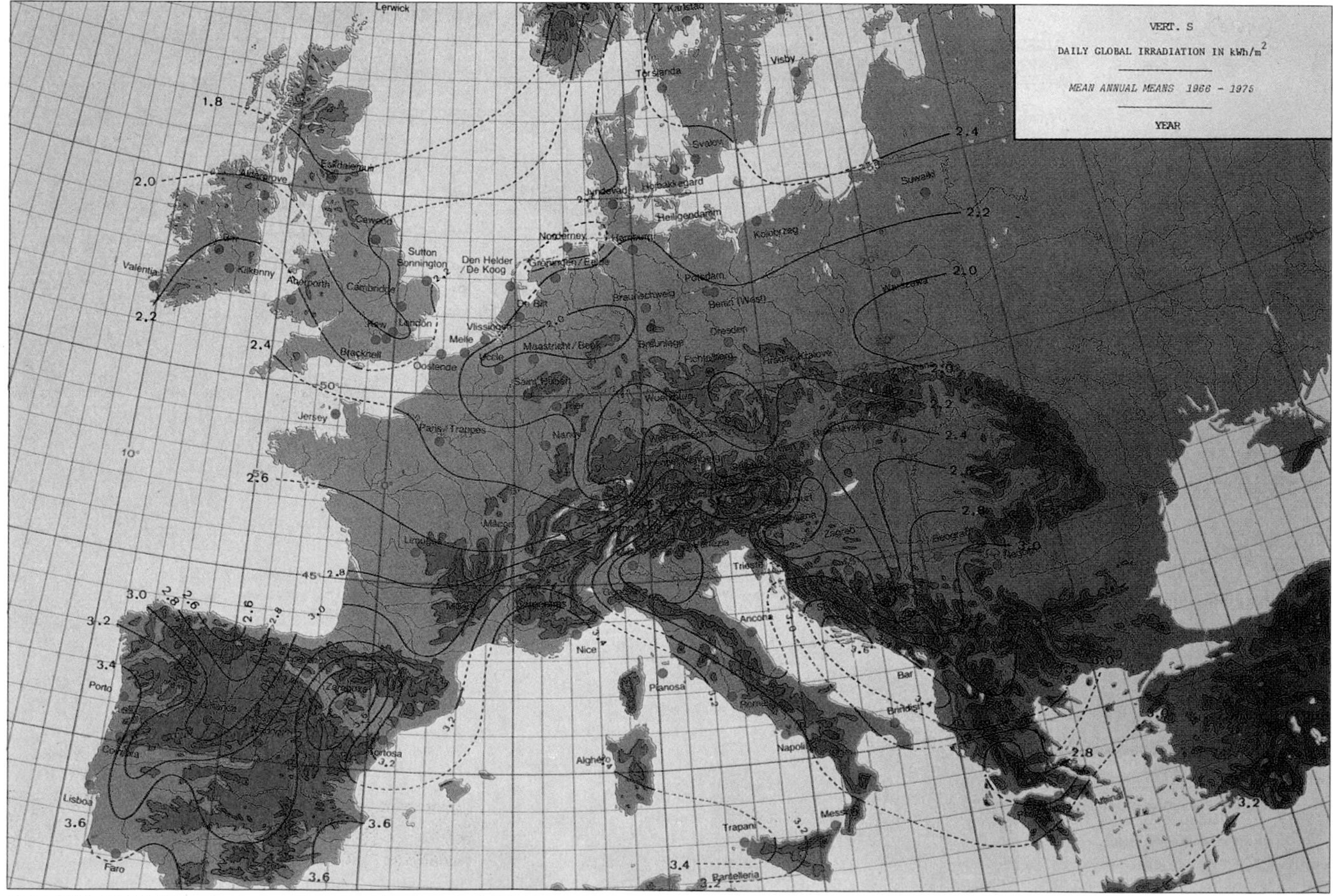

SOUTH ANNUAL Daily global irradiation in kWh/m2 : Mean annual means 1966 - 1975

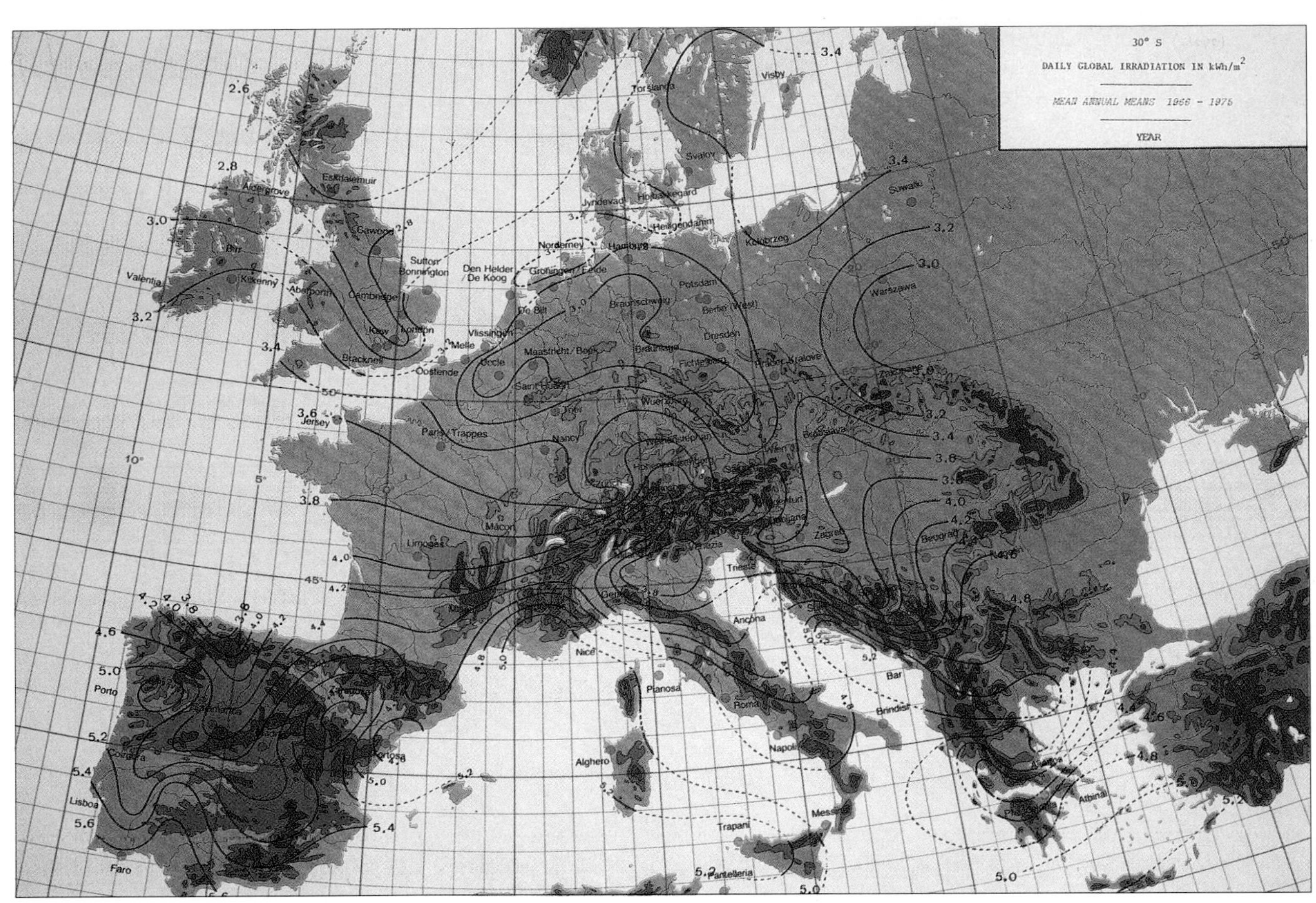

30° SOUTH Daily global irradiation in kWh/m² : Mean annual means 1966 - 1975

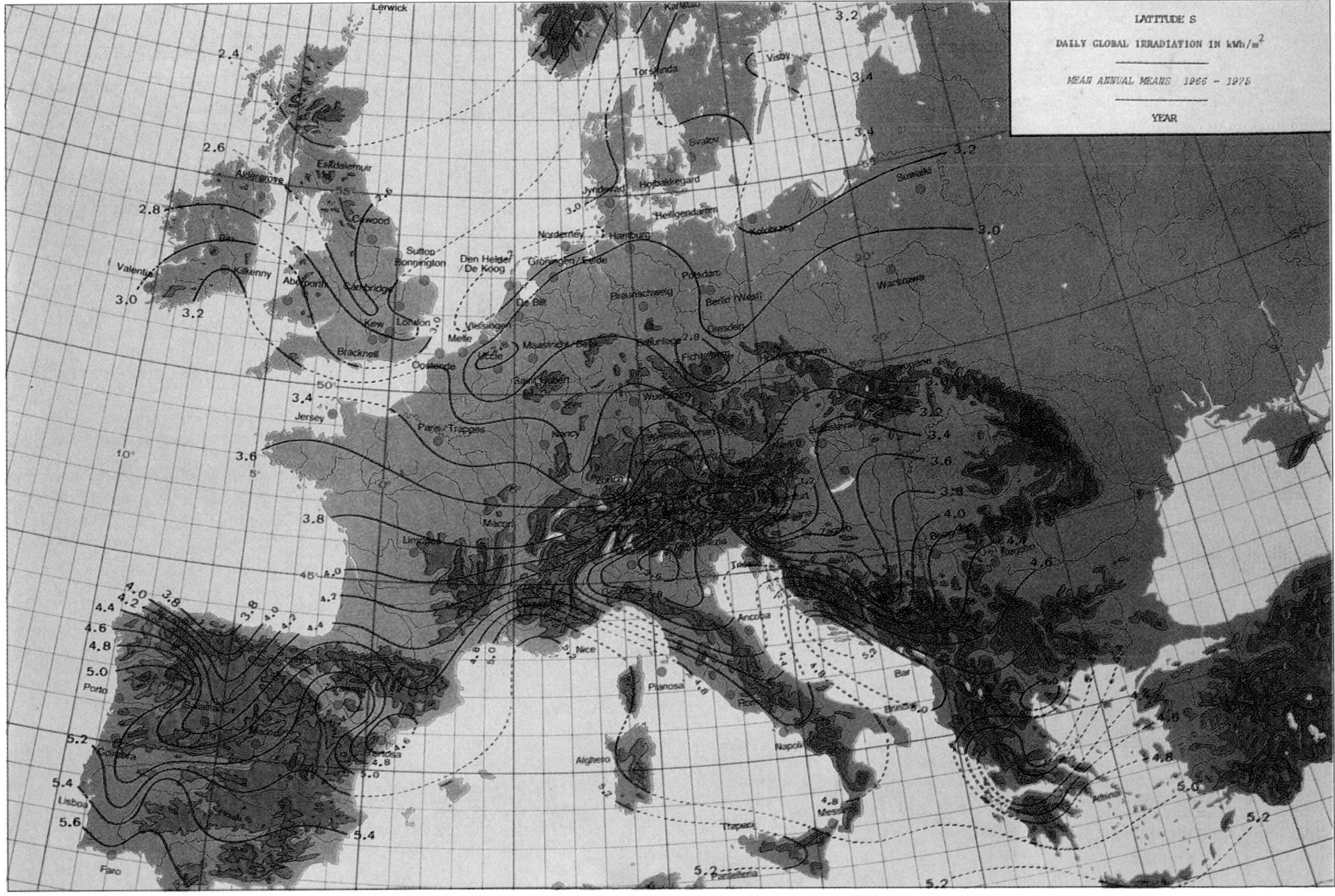

LATITUDE SOUTH Daily global irradiation in kWh/m² : Mean annual means 1966 - 1975

Diurnal variatons of temperature on clear and overcast days in winter and summer at Rye, Sussex, 1.1m above ground. Source Lacy.

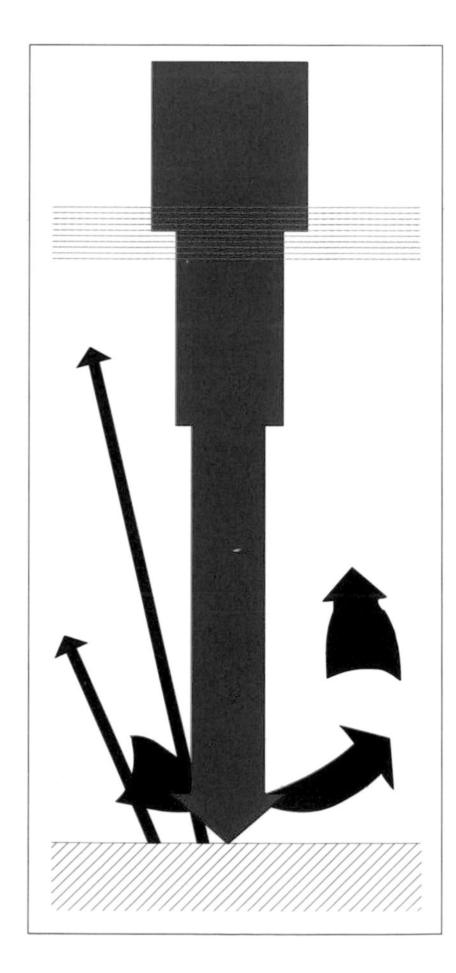

TEMPERATURE

At any instant, the air temperature at a site depends on two factors; incoming air flows driven by large-scale weather systems, and local climatic energy inputs. The latter modify the temperature of the incoming air to a greater or lesser extent, according to its speed. When the wind speed is slow, site factors such as the heating of the ground by sunshine and night-time cooling from outgoing long wave radiation exert a major influence on the air temperature close to the ground. With high wind speeds, the temperature of the incoming air mass is less affected by local factors.

Local inputs of climatic energy have a significant effect on the daily air temperature swing close to the ground. As one moves further up from the ground, the impact of ground diurnal temperature variations rapidly decreases. Therefore, under most meteorological conditions, mean daily temperature decreases with distance from the ground.

To achieve comparable measurements of temperatures at different sites, thermometers measuring air temperature are mounted at a standard height of about 1.2 m above the ground in a whitened, insulated and ventilated meteorological screen. These screens are normally set above short mown grass on level ground, well away from trees, buildings, walls and other obstructions. Temperatures measured closer to the ground will reveal a bigger range of daily variations.

The typical mean temperature of the incoming air mass depends on its place of origin. Air from the polar regions is normally cold and dry. Air from the Atlantic Ocean is usually humid and is relatively warm in winter and relatively cool in summer. Air from the east is typically cold and dry in winter and relatively hot and turbid in summer. The hot, dry and dusty air from the Sahara sometimes impinges on southern Europe and, occasionally, even further north.

In both summer and winter, directional temperature data for Hamburg can be regarded as representative of northern Europe.

The ground of the site is heated by incoming solar radiation. It is cooled by convection, long wave radiation and evaporation of water. The evaporation of ground water through irradiated vegetation - a process called evapotranspiration - is particularly important for controlling air temperatures. The highest temperatures are found in hot sunny weather over dark surfaces with no vegetation cover.

The heating effect of the ground on the air determines air temperatures at habitable levels. There is a daily temperature swing with maximum temperatures usually occurring in the afternoon and the lowest temperatures just after dawn. In overcast weather the daily temperature swing is usually small.

Very close to the ground, the temperature of the air approaches that of the ground surface. This influence decreases with distance from the surface. In the middle of a still night, the external air temperature at the top of a tall building may remain substantially above that at the ground floor.

At night, the surface temperature of the ground may fall below the air temperature because of emission of long wave radiation to the sky – a process which is greatest when there is no cloud. If the surface temperature falls below the dew point, then dew (or ice, if conditions are sufficiently cold) will form. Ground frosts are most likely to occur at low wind speeds.

When the ground is heated strongly, warm air begins to rise in quite large parcels. This ascending warm air is cooled and, if it is humid enough, will form convective cumulus clouds. In fine weather, the amount of cloud tends to increase towards the afternoon. The circulation of air by convection also increases wind movement in the afternoon. This effect dies down again in the evening.

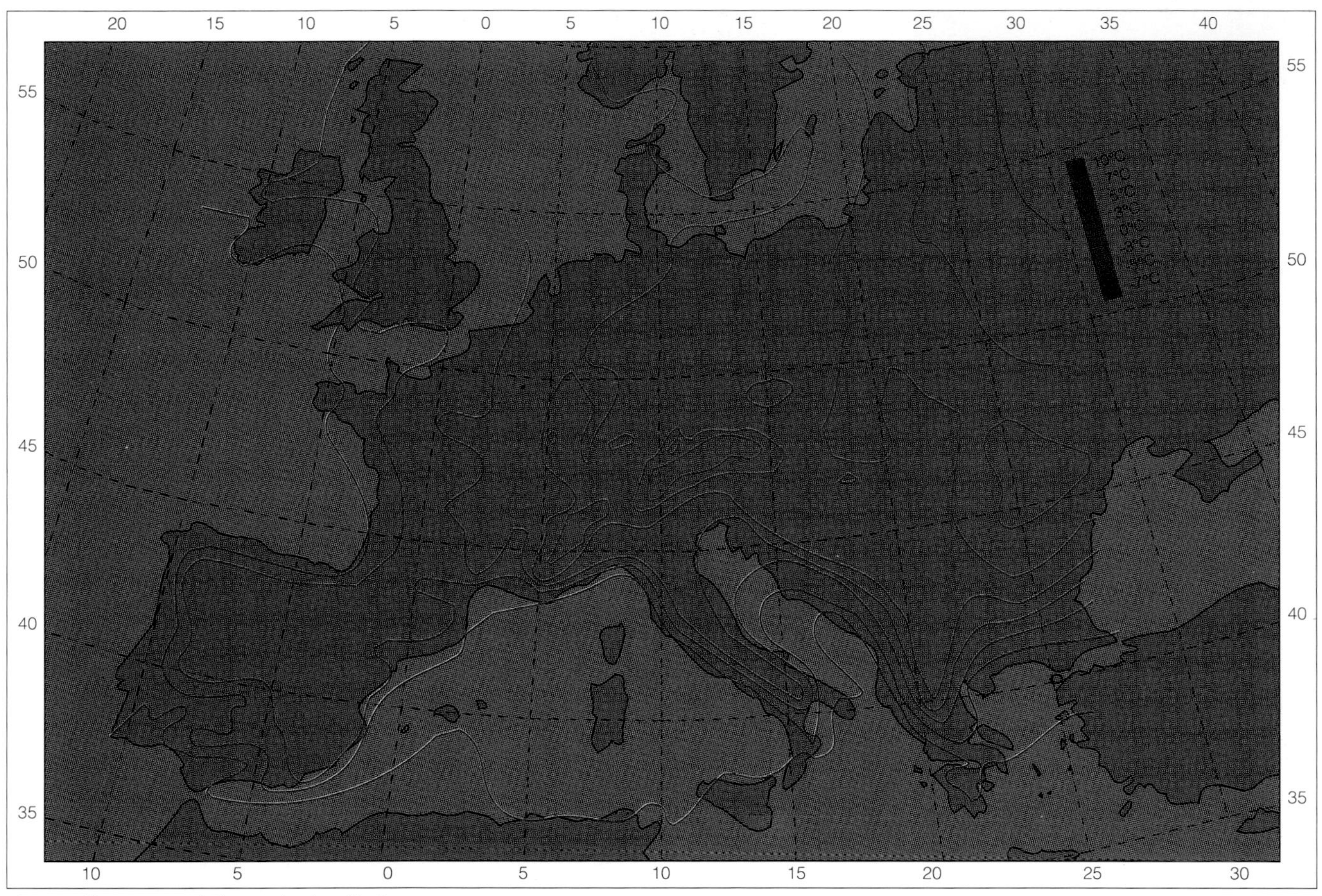

Average Temperature (c°) at station level in January.

Average temperature (C°) at station level in July.

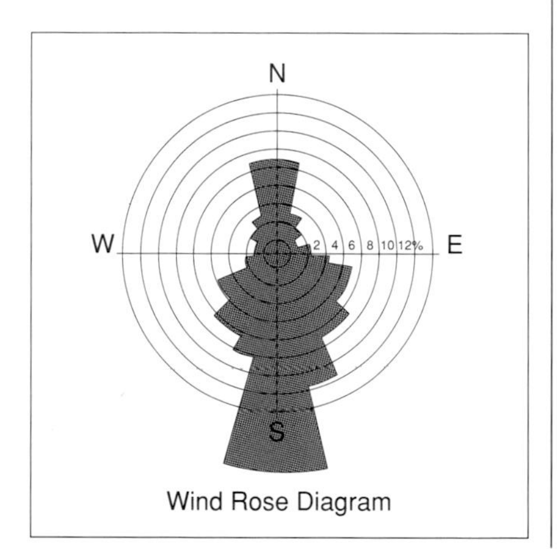

Wind Rose Diagram

WIND

Winds or air movement in the earth's atmosphere are caused by pressure differences generated by complex climatic factors. Wind speed and direction are normally measured in meteorological networks at a height of 10 m. Where possible, sites are selected which are exposed in all directions. As substantial obstacles to air flow exist at ground level in towns, urban air movement is generally quite turbulent.

For European regions north of the Alps, the prevailing winds are from the south-west. In winter these winds are generally warm and bring rain. Southern winds are warm and dry. Northern, polar winds are cold. North-eastern winds in winter are cold and dry. Around the Alps and Pyrenées, cold winds blow from the summits to the warmer lowlands.

Wind is a major design factor for architects. It affects comfort and influences rainfall. It modifies the heat exchange of a building envelope through convection and it causes infiltration of air into the building.

Wind resources at 50 metres above level for five different topographic conditions

	Sheltered terrain		Open plain		At a sea coast		Open sea		Hills and ridges	
	ms^{-1}	Wm^{-2}	ms^{-1}	Wm^{-2}	ms^{-1}	Wm^{-2}	ms^{-1}	Wm^{-2}	ms^{-1}	Wm^{-2}
	> 6.0	> 250	> 7.5	> 500	> 8.5	>700	> 9.0	> 800	> 11.5	>1800
	5.0-6.0	150-250	6.5-7.5	300-500	7.0-8.5	400-700	8.0-9.0	600-800	10.0-11.5	1200-1800
	4.5-5.0	100-150	5.5-6.5	200-300	6.0-7.0	250-400	7.0-8.0	400-600	8.5-10.0	700-1200
	3.5-4.5	50-100	4.5-5.5	100-200	5.0-6.0	150-250	5.5-7.0	200-400	7.0-8.5	400-700
	< 3.5	< 50	< 4.5	< 100	< 5.0	< 150	< 5.5	< 200	< 7.0	< 400

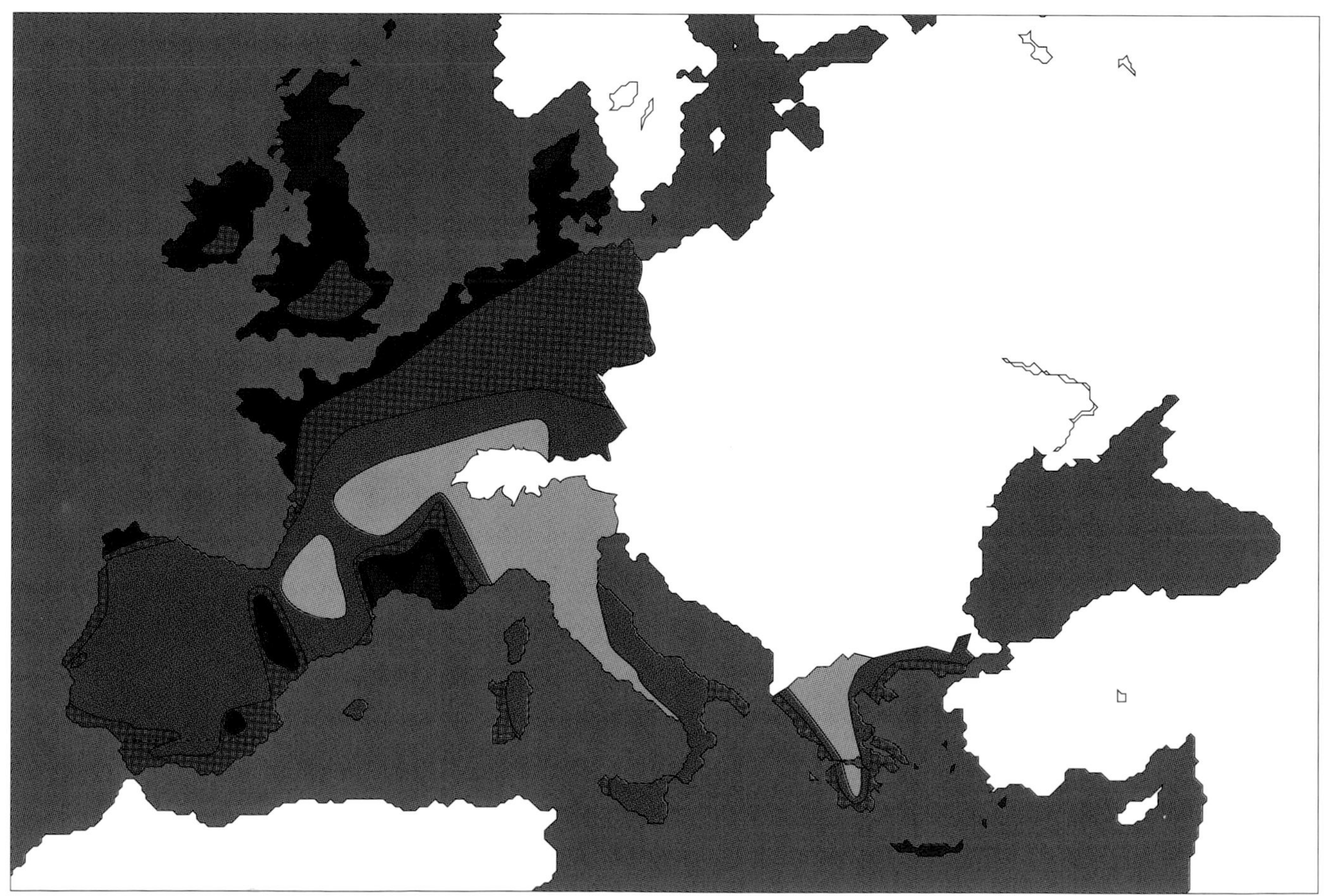

Distribution of wind resources in Europe. The Table shows wind energy at a height of 50 metres above ground level for five topographic conditions.

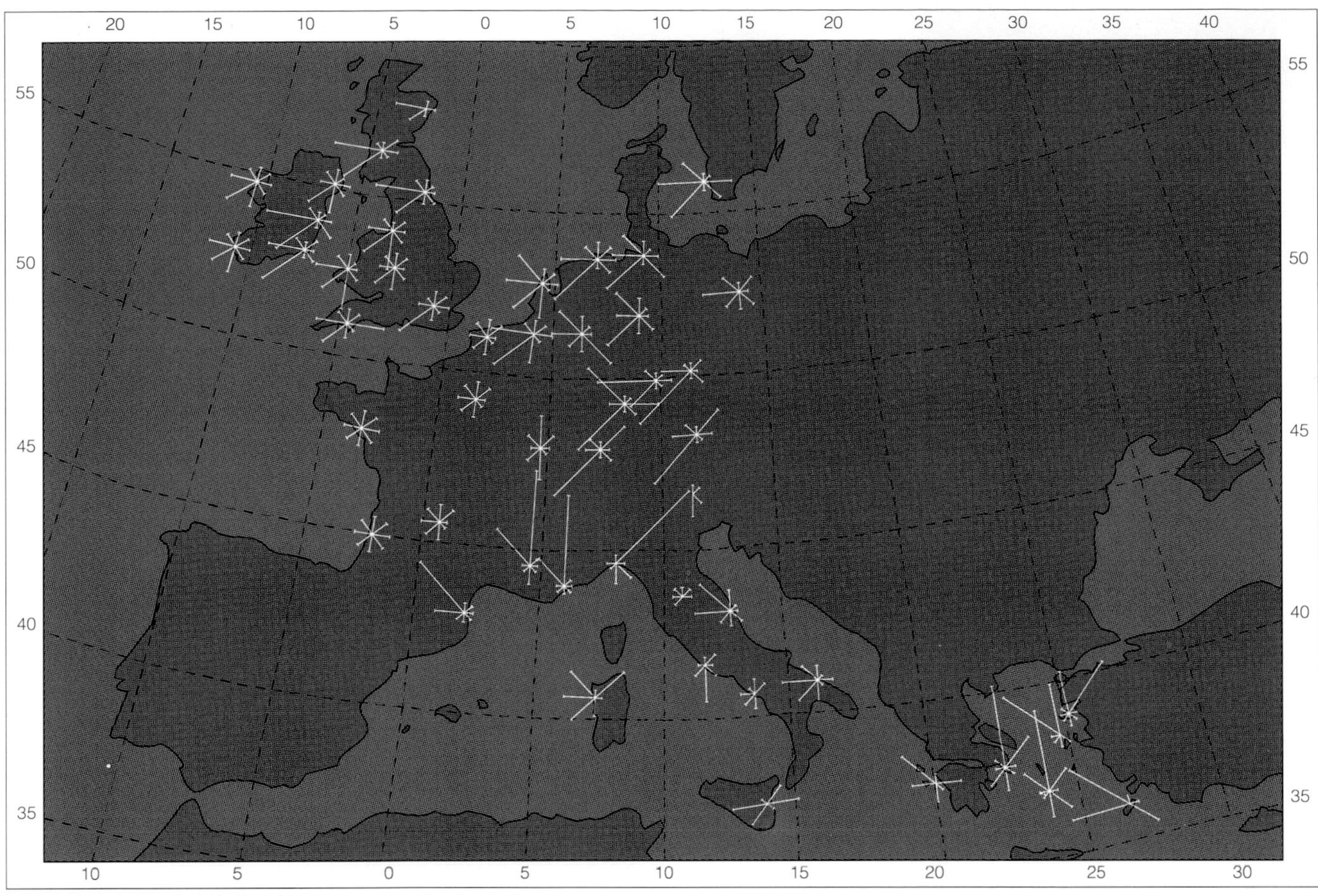

Annual frequency of wind direction.

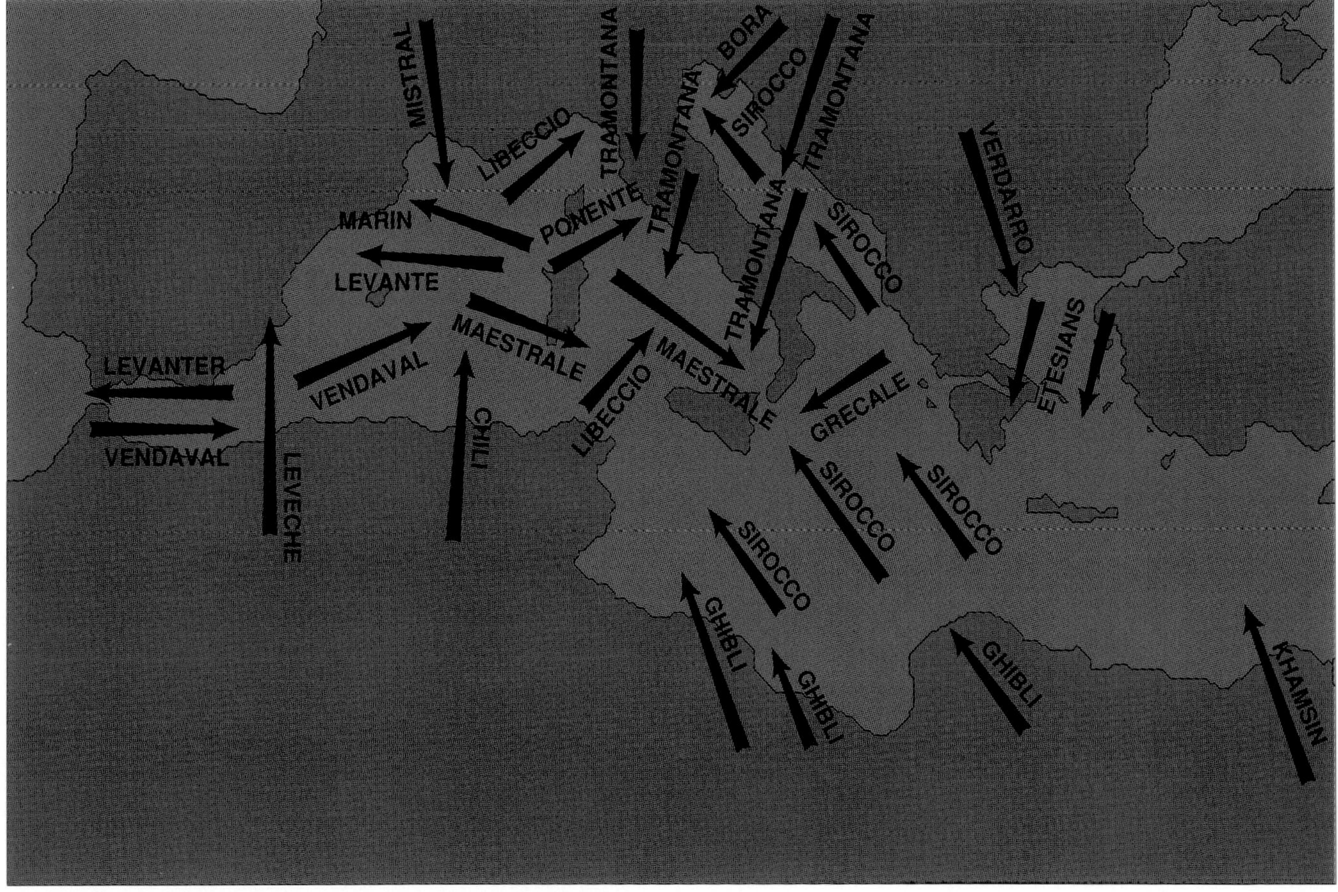

Principal winds of the mediterranean region (source European Wind Atlas).

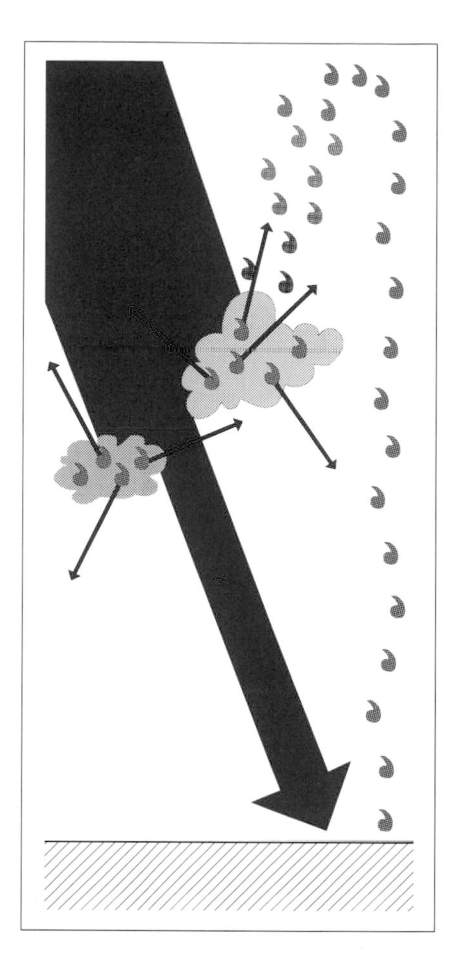

HUMIDITY

Air humidity may be described in four ways:
- dry and wet bulb temperature
- air temperature and relative humidity
- vapour pressure (in millibars)
- dew point.

The vapour pressure is the most stable variable across the day. The relative humidity varies considerably, tending to be highest close to dawn when the air temperature is lowest, and decreasing as the air temperature rises. This is because the relative humidity is related to saturated vapour pressure, i.e, the amount of water the air can hold at any given temperature. The saturated vapour pressure increases strongly with rise in temperature. The fall in relative humidity in the middle of the day tends to be greatest on summer days but such days remain uncomfortable because the vapour pressure does not fall. In winter, monthly mean daily relative humidities are very high in western Europe. They tend to increase as one proceeds towards the Atlantic Coast. In summer, they are lower but still quite high due to the Atlantic influence. They fall with dry winds from the polar regions and dry winds from the desert areas south of the Mediterranean.

When humid air is cooled to its dew point, the vapour will form dew, hoar frost, fog, rime ice or cloud droplets. The latter may coalesce to form rain, hail or snow.

In areas with high humidity levels, transmission of solar radiation is reduced because it is absorbed by water vapour and scattered and absorbed by clouds. Very dry air, on the other hand, causes hot days and cold nights.

In most parts of Europe, the humidity level is inside the comfort range for most of the year. Severe thermal discomfort only occurs when high vapour pressures combine with high temperatures to give hot, humid conditions or low humidity combines with high temperatures to produce hot, desiccating atmospheres. These conditions are most likely to be found near the Mediterranean Sea.

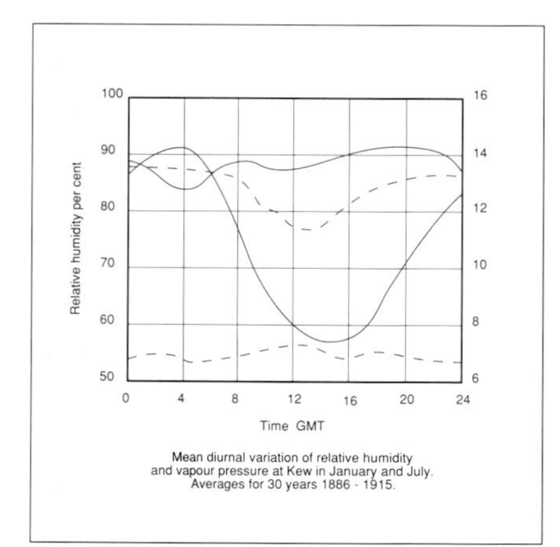

Mean diurnal variation of relative humidity and vapour pressure at Kew in January and July. Averages for 30 years 1886 - 1915.

CHAPTER 2

THE CLIMATE MESOCLIMATE

The climatic factors described in the preceding section on macroclimate are influenced by local conditions such as topography, vegetation and the nature of the area and its environs.

SOLAR RADIATION

Two factors have a major influence on the solar radiation received at a particular site: the turbidity of the atmosphere and the presence of geometric obstructions.

TURBIDITY

Turbidity consists of dust and suspended droplets of water, etc., which partly absorb and partly reflect (i.e. scatter) the solar radiation as it passes through the atmosphere.
Turbidity in the general continental atmosphere of western Europe is highest in summer when the amount of dust is greatest and large amounts of water vapour can produce very hazy skies.
In towns, pollutants from the concentration of cars, factories, heating systems, etc., absorb and scatter sunlight, weakening the direct solar beam but increasing the diffuse radiation on cloudless days. A dome of pollution can sometimes be seen above cities. In recent years, however, improvements in smoke control from factories and heating systems have lessened the turbidity difference between town and countryside. However, pollution from traffic has increased so that, overall pollution has become more widespread. The condition in towns is now at its worst in summer when sunlight acting on traffic exhausts produces some very unpleasant effects.
Low-level pollution by solid particles can be reduced by the presence of trees. The leaves act as a filter. Dust particles either cling to their surfaces or, having fallen on them, drop to the ground below. The air in the centre of an urban green space with plenty of trees is purer than the air near the perimeter.

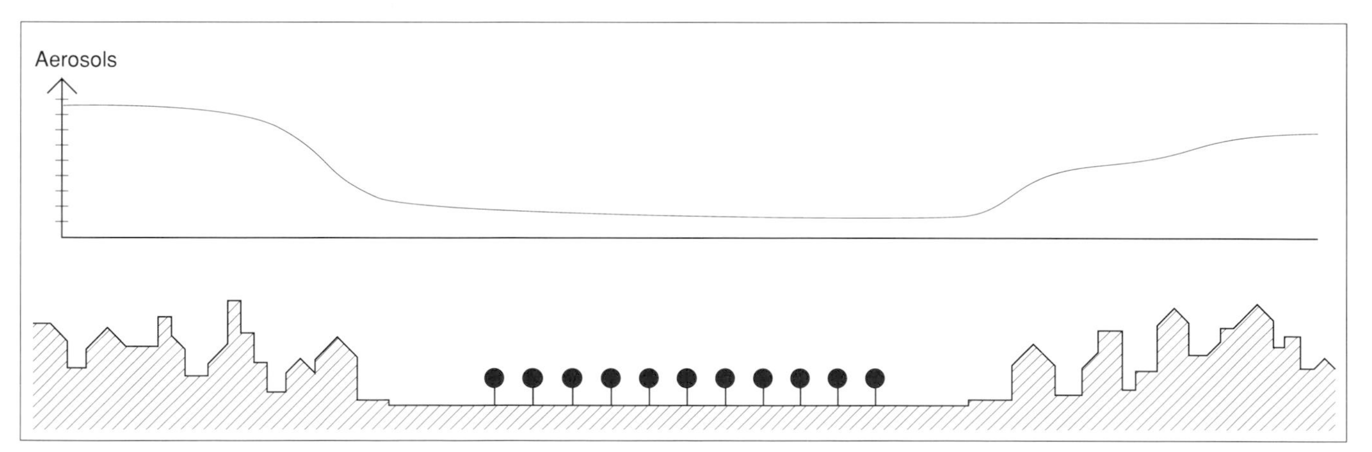

GEOMETRIC OBSTRUCTIONS

Geometric obstructions can be classified into three general classes – those related to the topography of the area, vegetation on or near the site and nearby buildings. They all, to a greater or lesser extent, shade the site from the sun. The precise impact of the obstructions on the amount of solar radiation received can be assessed geometrically using solar charts of the type described in the section on macroclimate, coupled with some ancillary aids.

TOPOGRAPHY

The geometrical assessment has to take into account the three-dimensional nature and seasonal effect of the surrounding terrain. Obstructions to the south tend to cause most overshadowing in winter because of the sun's low altitude. Valleys running east-west therefore face the greatest risk of permanent overshadowing from the southerly slope in winter. In northern Europe it is best, where possible, to locate buildings high enough up the south slope to catch significant amounts of winter sun.

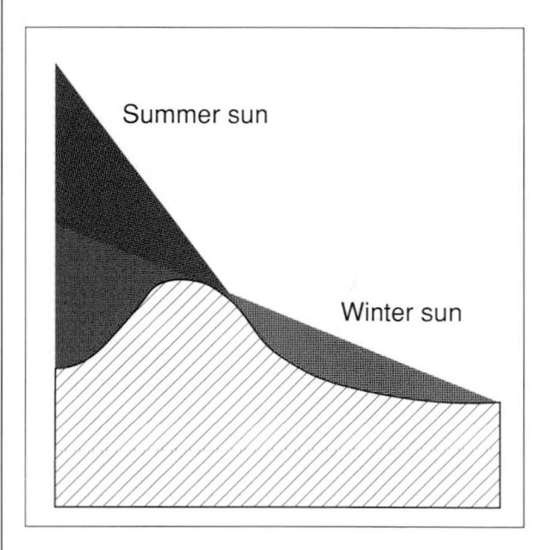

VEGETATION

The effect of deciduous vegetation also varies according to season. Overshadowing is diminished when the leaves fall in autumn. When the deciduous trees are in leaf, part of the incident sunlight diffuses through the leaves and the radiation is not blocked entirely.
Conifers, on the other hand, block sunlight to a greater extent throughout the year.

BUILDINGS

Surrounding buildings have an effect on the amount of sunlight and diffuse radiation received at the site. Their impact on sunlight availability changes with season and it is necessary to bear this in mind in developing a site. Not only must the effect of existing construction be taken into account in carrying out the assessment but also the influence of likely future building developments.

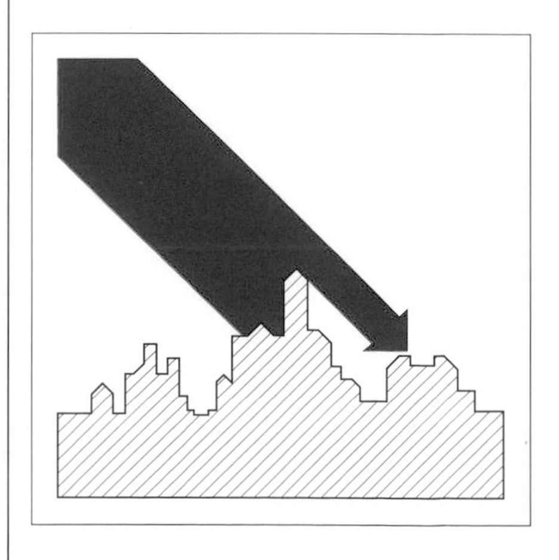

TEMPERATURE

The temperature of the air at a site is influenced by topography, vegetation and the nature of the nearby ground surfaces.

TOPOGRAPHY

Topography influences air temperature because of its effect on orientation and tilt of the ground, wind exposure, night-time cooling and flow of heated and cooling air.
Ground surfaces oriented and tilted towards the sun are more strongly irradiated than other surfaces. When the sun is shining, favourably-inclined surfaces become warmer compared with unfavourably oriented, untilted and/or overshadowed surfaces.
In sunny weather, surfaces most exposed to wind will experience the smallest temperature rises: the wind will rapidly remove surface heat by forced convection, substantially reducing any potential warming. This effect can be considerable in, say, a hill site when the sun is in the south-west and the wind in the north-east.
Under night-time cooling conditions, if the air outside is hot its capacity for cooling the building surfaces will be reduced.
The various flows of heated rising air and cooling sinking air will be affected by the structure of the terrain changing the temperature patterns. In a complex terrain this can produce a very wide range of microclimates. For instance, on sunny days, valleys are generally warmer than hilltops. At night, however, as the slopes cool the air in contact with them runs down into the valley to form pools of cool air at the bottom. Thus, at night sites on favourably-oriented slopes may be warmer than those in the valley.

VEGETATION

In well-wooded areas, the trees intercept 60% to 90% of the solar radiation causing a substantial reduction in the daytime increase of the surface temperature of the ground below. The air below foliage remains cooler than elsewhere. This produces a stable configuration of layers of cooler (heavier) air below warm (lighter) air masses round sunlit foliage. As a result, there is a reduced turbulence and air movements in the layers close to the ground. This daytime phenomenon may be permanent or seasonal depending on whether the trees are deciduous or evergreen. After dark, the foliage hinders the outgoing long wave radiation and reduces the night-time temperature drop. Diurnal temperature differences are therefore smaller in woodlands than in open countryside.

GROUND SURFACES

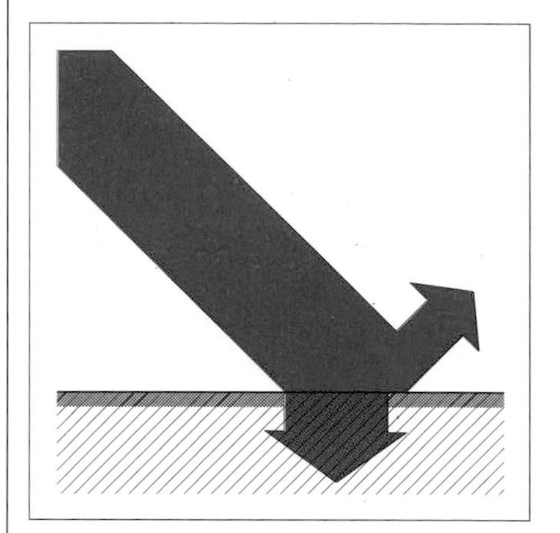

The temperature of the air is influenced by the nature of surrounding surfaces which intercept the solar radiation. The colour of the ground affects the relative proportions of incident radiation which are absorbed and reflected; dark colours tend to produce high surface temperatures. Other ground surface properties also have an effect on air temperature. In considering this whole subject it is useful to categorize ground cover into three general types – vegetation-covered areas, surfaces covered with dry materials such as concrete, brick, etc., and surfaces covered with water.

Lawns and areas covered with low shrubs are examples of vegetation-covered areas where surface temperature cooling takes place by evaporation of the water transpired through the leaves. As the surfaces of the leaves do not heat up much in the sun this process reduces the air temperature above the vegetation throughout the day. It does, however, increase vapour pressure.

Concrete, bricks, gravel, cobble stones and other materials with a high thermal inertia, when set in a layer over the underlying earth, are all examples of dry ground cover. The temperature rise of these surfaces depends on the surface colour. Heat is stored in the day and re-emitted in the evening. This emission of radiant heat can be very noticeable under the still conditions which often occur in hot weather.

Lakes and ponds can easily store considerable amounts of heat for a relatively small temperature rise. Because water bodies do not heat up very much when subjected to radiation during the day nor cool down very much at night, they act as thermal regulators. The stable surface temperature influences the temperature of the adjacent air, producing cooler temperatures in the daytime and warmer ones at night.

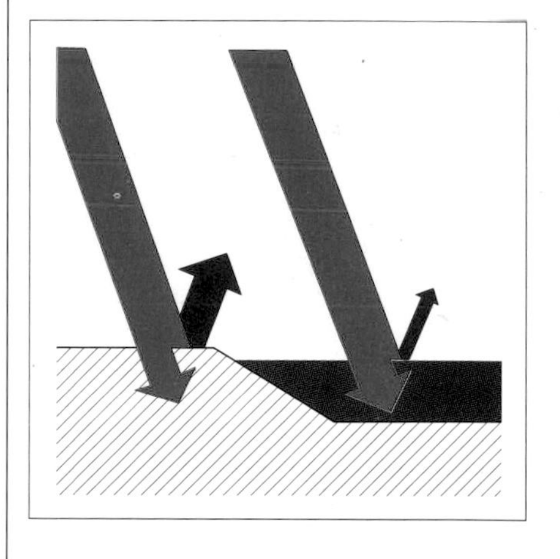

Ground surface cover has a noticeable effect on air temperature in towns. Invariably, there are a lot of roads constructed of heavy building materials. Rain water is usually led away rapidly. The amount of vegetation cover is often small. Therefore, there is little potential for evaporative cooling. In addition, there are considerable heat inputs from vehicles, factories, heating plants, etc. All these combustion processes affect the atmosphere, decreasing solar radiation. The pollution dome alters the long wave radiation transfer. Large towns, therefore, tend to be quite a bit warmer than the surrounding countryside for most of the day. The difference is especially marked in still weather in late evening. In the morning, towns heat up less rapidly on account of their large thermal inertia. The precise extent of the town-country difference depends, of course, on the size of the town. For a large town, the typical daily mean difference is 1-2 degrees C. The peak difference on a still evening is much greater. Temperatures may vary by 5 to 10° C between densely built areas and city parks

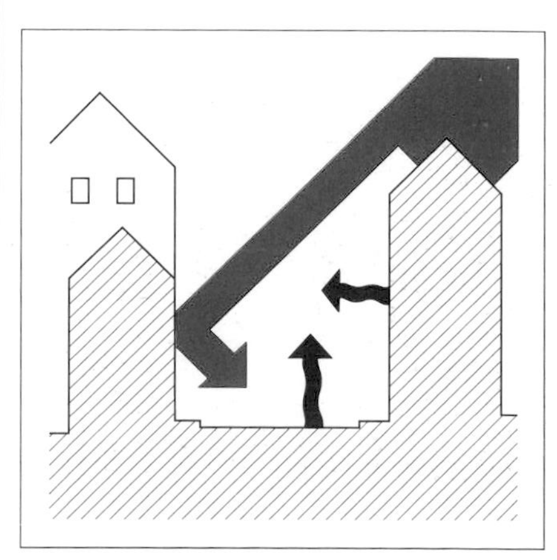

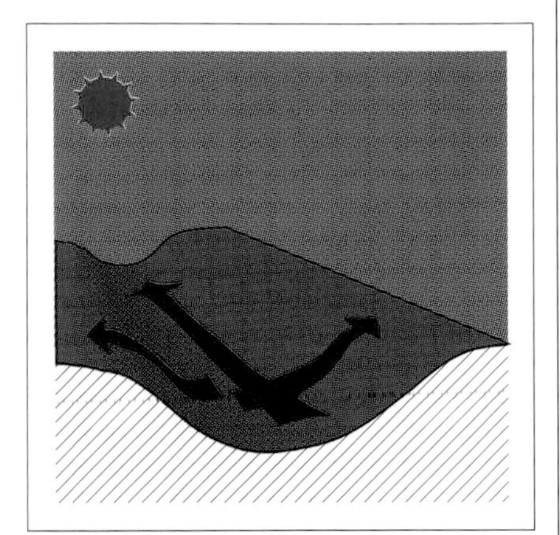

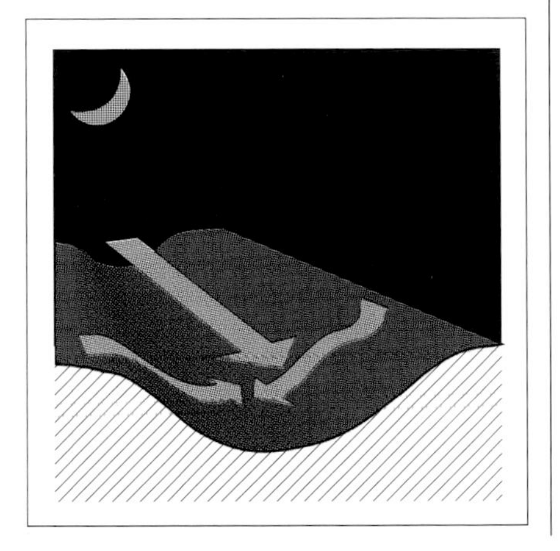

WIND

The town-country differences described in the preceding section can have an effect on the air movement experienced at a particular site. Wind flow is also influenced by topography.

TOPOGRAPHY

The terrain of an area can cause medium or large-scale modifications to wind flow at a site. For instance, topographical features can provide protection for certain sites but at the same time over-expose others. They can also modify the direction of the prevailing winds over considerable areas. The wind flow at the crest of a hill can be accelerated because of compression of the air streams.

Air in contact with surfaces warmed by solar radiation tends to rise while air in contact with cold surfaces (for example, those experiencing night-time radiative cooling) tends to sink. The resulting density changes generate air flow patterns which are characteristic of the particular terrains involved. Several terrain configurations cause cyclical air flows. Examples are water-land interfaces, hillsides and valleys.

With a water-land interface, the lake surface is warmer than the adjacent land in winter. On calm days in winter, therefore, the air tends to flow from the land to the water. On summer days, however, the land surface is warmer than the water surface and the direction of flow is reversed. In addition diurnal effects come into play in summer.

During the afternoon, the land can be so much hotter than the water that a breeze off the lake develops. At night, the water surface may not cool down as much as the land so the air movement is in the opposite direction.

On hillsides, solar radiation can increase surface temperatures and the hot surfaces generate uphill surface-air streams during the day. In mountain regions this phenomenon is known as an up-valley breeze. At night, when the surfaces are no longer receiving solar radiation, they begin to cool down. The temperature gradient decreases and finally reverses and the air circulates in the opposite direction. This is a down-mountain breeze.

In long valleys these phenomena tend to create length-wise air movements so that the longer the valley and the higher the surface temperature, the stronger the air flow. Complex air movements can result from a combination of the valley effect and the hillside effect.

TOWN-COUNTRY TEMPERATURE DIFFERENCES

In still weather, the temperature in towns is higher than that in the surrounding open country for a significant part of the day. As a result, wind flows can be generated which are similar to those found when warm masses of water are adjacent to cooler land. The urban heat island of warm air can cause a flow of wind towards the town centre.
Similar flows can occur within towns, from urban spaces, such as parks, towards adjacent buildings.

HUMIDITY

The topography of an area and the presence of vegetation both have an effect on humidity.

TOPOGRAPHY

Topographical features can force the water from precipitation to flow preferentially towards hollows in the ground and create humid soil pockets. In still, sunny weather the air above these pockets is cooler than the air above adjacent dry ground.

The presence of lakes, rivers and seas also has an effect on humidity. As part of the evaporation process, sensible heat is extracted from the air close to these water surfaces and the air becomes cooler and more dense as a result. Provided the vapour pressure of the cooled air remains within an acceptable range, this process can aid summer-time comfort.

VEGETATION

In sunny weather, the air close to the ground is cooled by the transpiration of water through the foliage of trees and blades of grass. The rate of transpiration drops in cloudy weather.
During one week of sunshine in Germany for example, one square metre of grass will evaporate about 20 litres of water.

TYPES OF MESOCLIMATE

COASTAL REGIONS

Along the coast, the sea has a modifying effect on the daily temperature variations found further inland. On clear winter days, for instance, the air temperature at the coast is higher than that further inland whereas in summer it is cooler and more humid.
The absence of obstacles like trees and buildings and the low surface friction over the sea causes winds off-shore to be considerably stronger than those inland. In moderate sunny weather when the land is warmer than the sea, a sea breeze can arise which blows from the sea towards the land This is most likely to occur in the afternoon and can be a significant feature in a coastal area. It is, for instance, often found in spring and summer on the north-west coast of Europe. Such winds tend to reverse direction at night.

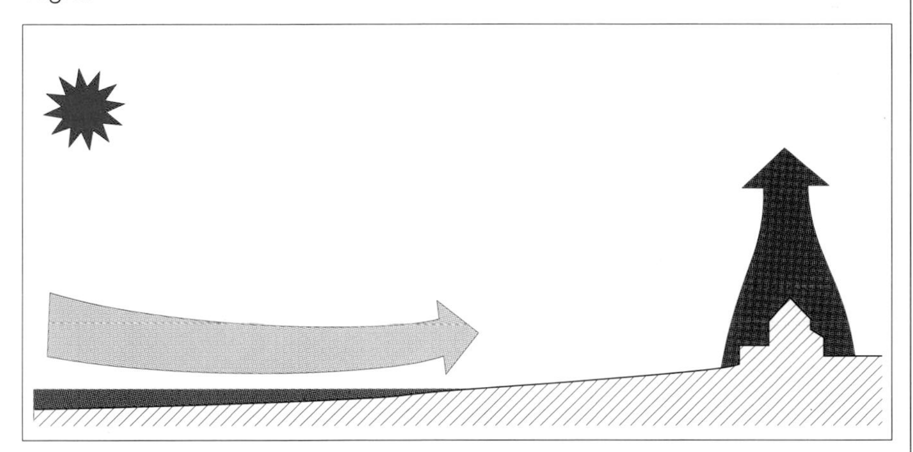

FLAT OPEN COUNTRY

In flat, open country there will be few major obstacles. Those most likely to exist will be hedges, lines of trees and nearby woods or villages.
Solar radiation conditions in such areas are likely to be close to the mean macroclimatic data for the region.
Wind speeds will probably be above average because of the lack of obstacles. The nature of the ground cover has an important effect on wind in open country. Rough surfaces such as scrubland and hedgerows slow down the wind close to the ground more than do smooth surfaces such as short grass or stretches of water.
In open sites, the direction of the wind is similar to that found at the local meteorological station, provided the latter is itself on unobstructed terrain.

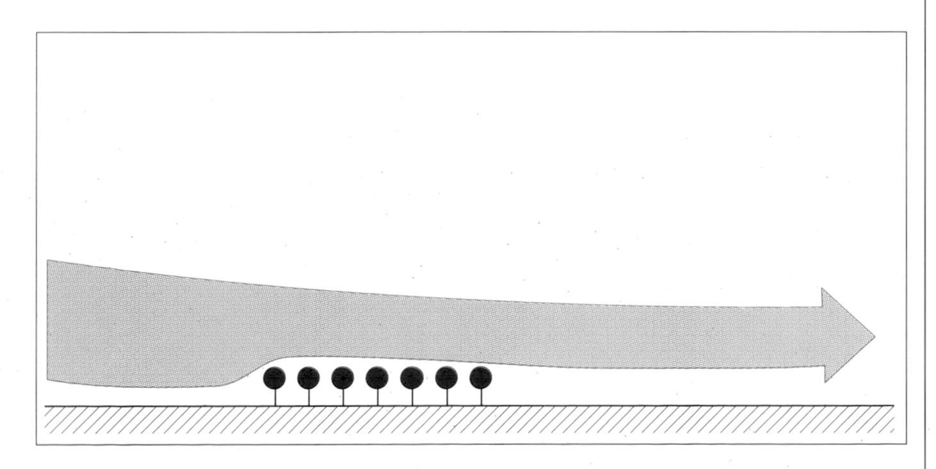

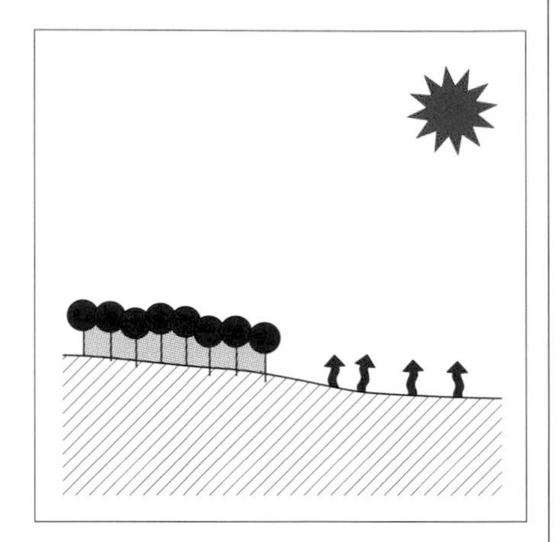

WOODLANDS

Trees within woods and forests constitute a screen for both sun and wind. In woodland areas apart from the clearings, shade is plentiful and winds are weak. In the day, the temperature in the underwood is lower than in open sites. The cool air masses remain in a stable position under the warm air masses and this tends to reduce even further the small amount of air flow which might be present. At night, the trees hinder the outgoing long wave radiation and this, coupled with the low air movement below the canopy, causes the night-time temperature in woods to remain higher than elsewhere.

VALLEYS

The orientation of a valley has a great bearing on its mesoclimate. If a valley is oriented in the direction of the prevailing wind, the air flow may be channelled strongly along the valley bottom. By contrast, a valley which runs perpendicular to the wind flow has its bottom and lower slopes well-protected from the wind above.

As far as solar radiation is concerned, unobstructed slopes lying between south-east and south-west are well-exposed. Slopes between north-east and north-west, on the other hand, do not receive much direct solar radiation. On such slopes, the sun's beam may in fact be totally blocked by the crest above.

The combined influence of sun and wind can have a big impact on temperatures at individual sites.

Any accumulated water in the valley bottom has a modifying effect on the daily temperature swings. It will also increase the humidity of the air or any wind flows streaming through the valley.

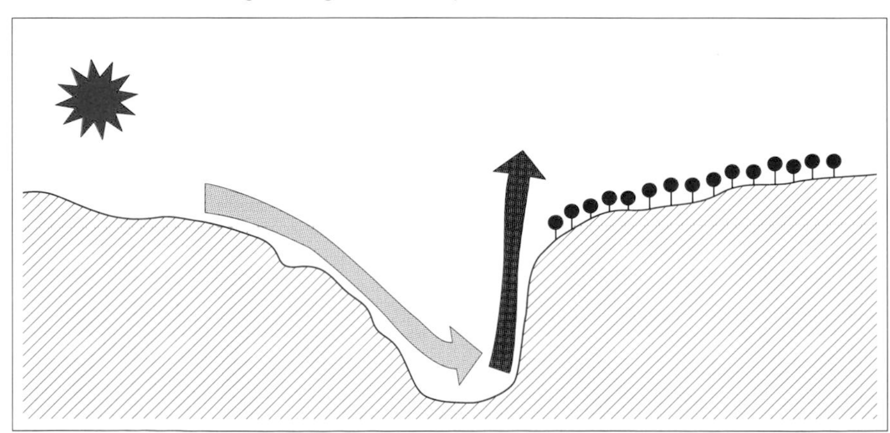

CITIES

When the regional winds are weak, the relative warmth of a large city compared to the surrounding areas can produce a convective circulation of air whereby warm air in the city centre rises and is replaced by cooler, more dense air flowing in from the countryside.

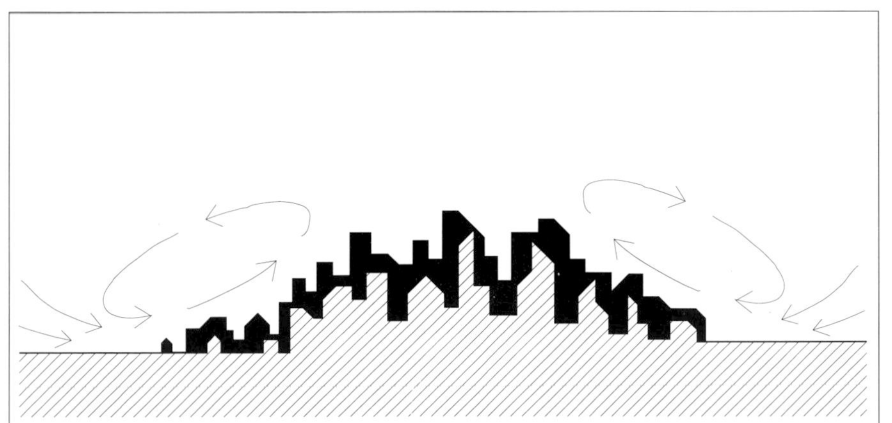

MOUNTAIN REGIONS

Climatic conditions in mountainous areas are significantly different from those in nearby flat open terrain. Because exposure to solar radiation and air movement are both dictated by the topography, each mountain slope has different climatic characteristics. The typical drop in temperature due to altitude may be about 0.7 degrees C for each 100 metre rise, although other factors may alter this. Similarly, a decrease in pressure of 1 millibar may typically be experienced for every 15 metre rise measured near an altitude of 2000 m.

Rain and snow are more frequent in mountain regions than in adjacent flat country. The increased rainfall is brought about when wind rises up a slope and the decrease in atmospheric pressure experienced with the rise produces cooling by expansion. This, in turn, causes some of the water in the air to condense. The rain often turns to snow at higher altitudes. Wind-exposed slopes are much more likely to experience rain than slopes which face down-wind.

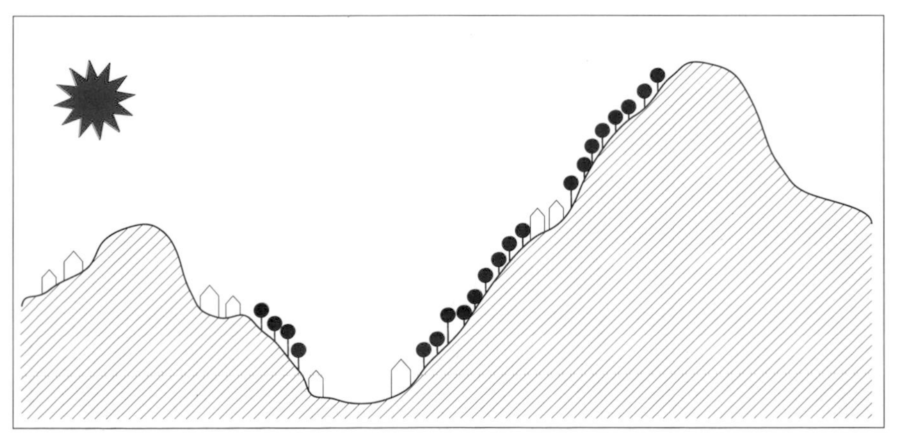

CHAPTER 3

THE CLIMATE
MICROCLIMATE

At the scale of the site, man's intervention can modify the environment close to buildings, creating conditions known as the microclimate or the climate of a small area.

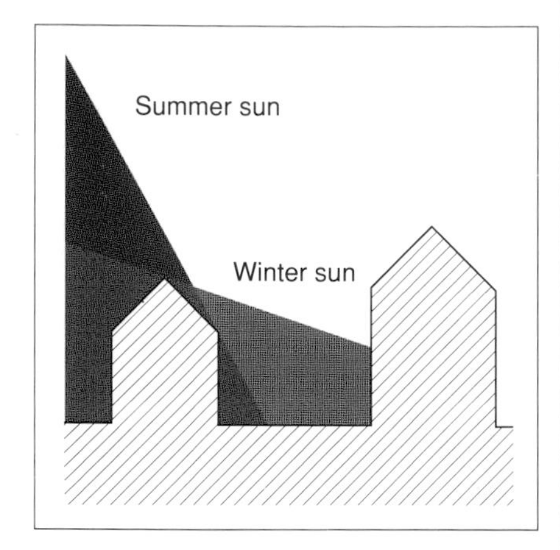

SOLAR RADIATION

The amount of solar radiation received at a site is dependent on local planting of vegetation and the shape, size and position of nearby buildings.

VEGETATION

Vegetation is different from other obstructions which intercept the solar radiation falling on a site. Certain types of planting change with the seasons. Many (deciduous trees, for example) provide only partial screening, filtering the incident radiation rather than blocking it completely, which can be used to advantage.

NEARBY BUILDINGS

Existing and potential future buildings in the neighbourhood of the site provide a fixed screen which must be taken into account in building design, especially in towns.

The solar altitude and azimuth for the whole year, hour-by-hour, can be plotted on a solar chart. The altitude scale is shown on a series of concentric circles. The azimuth scale is set around the perimeter of the chart. The azimuth angle is read by setting a straight edge from the centre of the chart to the intersection of the required hour and date path lines and noting where it cuts the chart perimeter. Different charts are required for different latitudes.

HUMIDITY

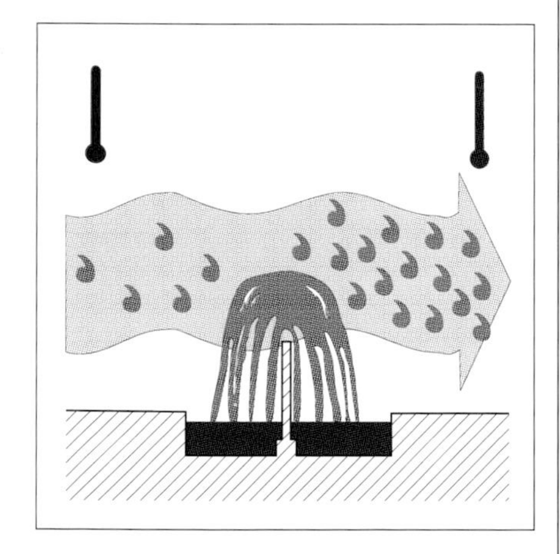

The humidity of the air at a site is modified by the presence of water and vegetation.

WATER

Fountains, water circulating under porous pavements, ponds and canals all bring about humidification – and hence cooling – of the adjacent air, although it is important to ensure that the humidity at a site remains within the comfort range.

VEGETATION

The evapotranspiration process of nearby vegetation also has a cooling effect on the air.

WIND

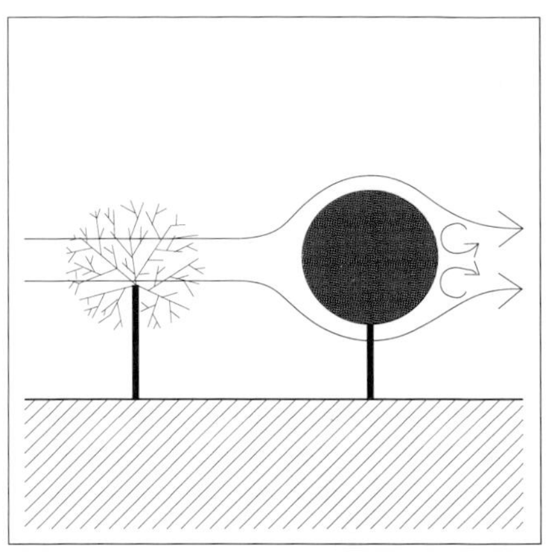

Local wind conditions can be modified by the presence of vegetation, buildings and built screens.

VEGETATION

Shelter belts are a common way of providing protection from the wind. Conifers offer year-round protection but obstruct sunlight in winter. Deciduous trees afford more shelter when they are in leaf in summer than when they are bare in winter. Even in winter, however, the bare branches still cause some reduction of wind speeds.

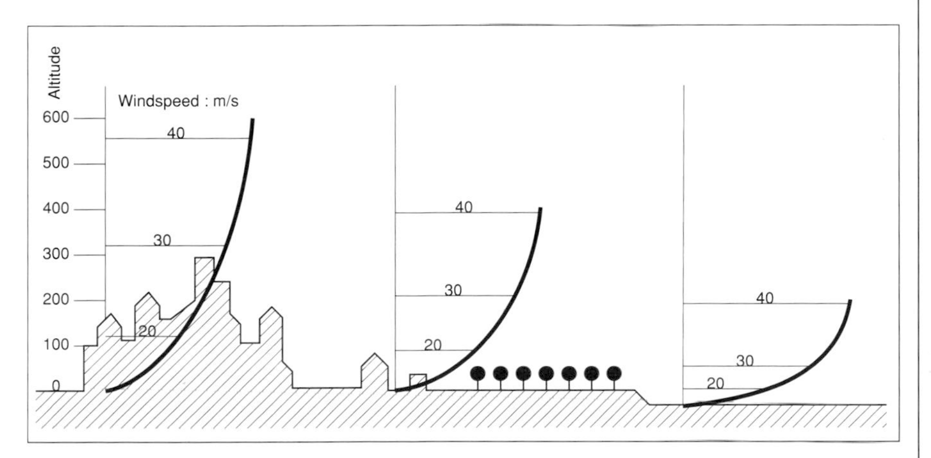

BUILDINGS

When wind encounters an obstacle its speed and direction are modified. A solid mass such as a building, forces the wind to go round or over it. The building side exposed to the wind is under positive (increased) pressure whereas the opposite, sheltered side experiences reduced pressure.
In general, wind speed increases with height above ground. Because of the number of obstacles to flow encountered in towns, the mean wind speed at a given height is lower in towns than over open land. The size of the obstacles influences the vertical gradient.
Wind flow in towns is more turbulent and changeable in direction than in the surrounding countryside. Particularly strong gusts may be experienced at the bottom of tall buildings.

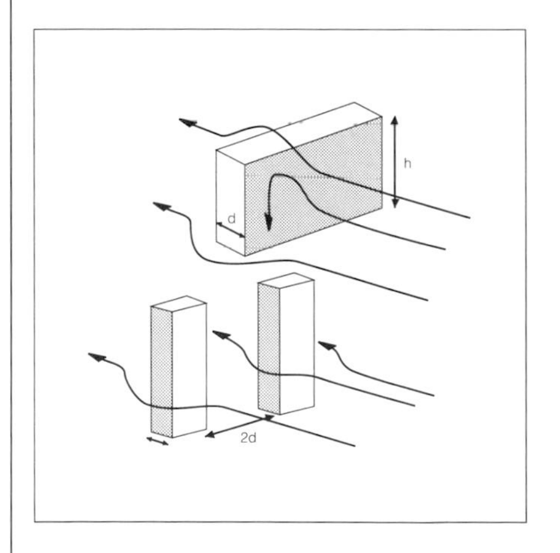

BUILT SCREENS

Deliberate sheltering can be created by planting as seen earlier, or by means of built screens. Nearby buildings, of course, can also provide shelter from the wind.

The sheltering efficiency of a long linear screen is determined by its height and permeability to wind. Close to the screen, dense screens create a larger wind speed reduction than do permeable screens. The depth of the sheltered zone, however, is not so great with a dense screen.

The depth of a protected zone is proportional to the height of the screen. For a screen of limited width such as a building, the sheltered zone increases in depth from the corners to the middle. The depth of the protected zone increases with building width until the latter is about ten times the building height. At this point the depth of the sheltered zone is about eight times the building height.

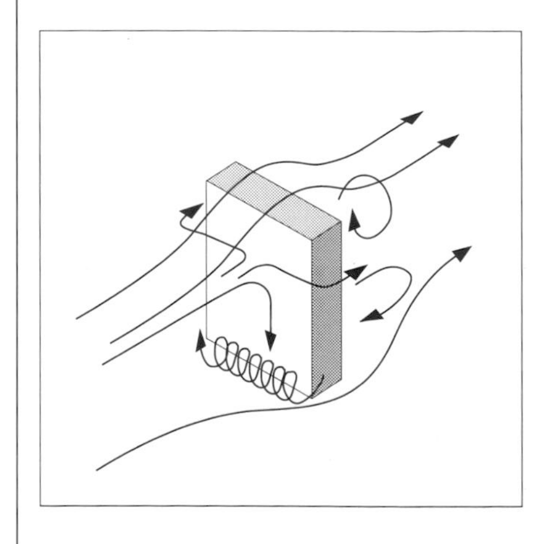

URBAN MICROCLIMATES

Urban microclimates are particularly complex because of the number and diversity of factors which come into play. Solar radiation, temperature and wind conditions can vary significantly according to topography and local surroundings. In addition, layout density can provide further constraints: the precise plot division, the need for access and privacy, and the noise and impact of atmospheric pollution must all be taken into account.

For any given project, specific design decisions need to be made on the basis of the microclimatic features which will have the greatest impact on the site.

In winter, most urban microclimates are more moderate than those found in suburban or rural areas. They are characterized by slightly higher temperatures and, away from tall buildings, weaker winds. During the day, wide streets, squares and non-planted areas are the warmest parts of a town. At night, the narrow streets have higher temperatures than the rest of the city. In summer, green spaces are particularly useful in modifying the environment during the late afternoon, when the buildings are very hot inside.

Strong local winds can modify the temperature distribution described above. Usually winds in towns are moderate because of the number and range of obstacles they face. However, a few configurations such as long straight avenues or multi-storey buildings can cause significant air circulation. Tall buildings rising above low-rise buildings can create strong turbulent wind conditions on the ground as the air is brought down from high levels. Strong winds can flow through gaps at the base of tall buildings. To protect pedestrians from this, the turbulent flow has to be prevented from descending to street level, for example by modifying the building form or by using wide protective canopies.
In semi-open areas, adjacent buildings can be used as protective screens against some winds.

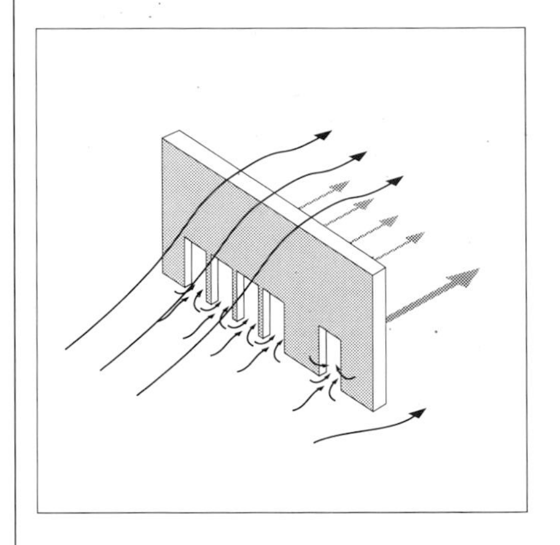

The strength of the solar beam in urban areas will depend on the level of particle pollution in the air streams above the town. Some pollutants - sulphur dioxide is an example - have no effect on beam strength.

The amount of solar radiation received at a particular site often depends on the shade cast by nearby buildings. The term 'street effect' is used to characterize the masking effect caused by buildings located across the street. The street effect depends on the height of the buildings and the distance between them as well as site latitude and street orientation and is expressed as a percentage of the usable solar gains.

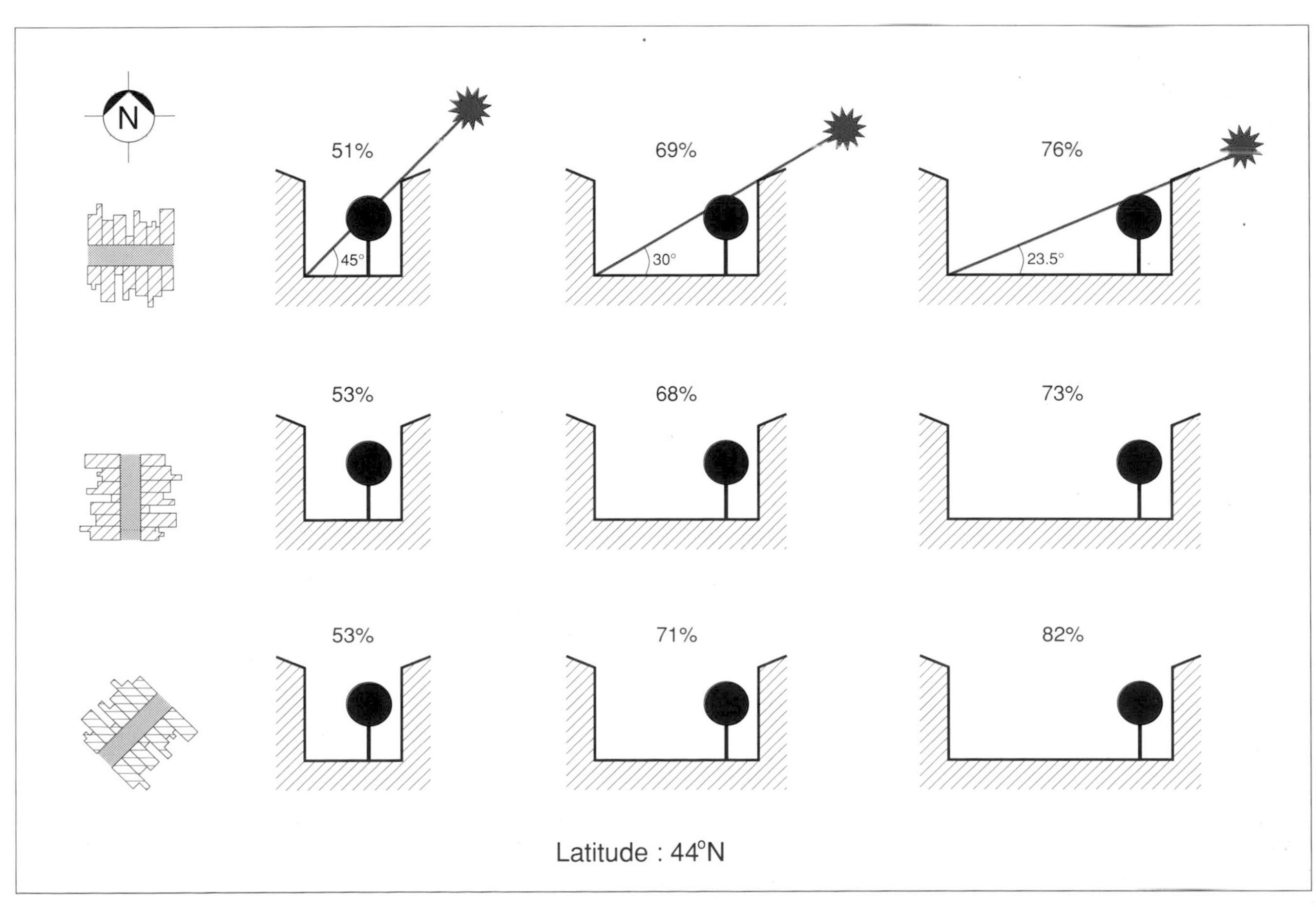

CONCLUSION

The overall objective of climate-responsive architecture may be expressed as the provision of high standards of thermal and visual comfort within and around buildings of quality which are energy efficient in use and also in their construction. It follows that the building should respond to the environment in which it is to be built in order to take full advantage of the useful climatic effects occurring on the site and that any undesirable conditions should be minimised or eliminated.

A knowledge of the regional and local climate is an essential input which will form the basis of analyses to characterise year-round performance. With this information, and ideally in addition some reliable local experience, it is then possible to appreciate 'positive' and 'negative' climatic influences which can be modified to improve the microclimate of the site around the building.

While the general macroclimate and mesoclimate of the region is beyond our influence, design changes at the microclimatic level can provide significant benefits.

This approach can help to minimise or even to avoid what are often more complex and expensive measures in the design of the building itself, and in addition can improve the amenity and extend the utility of outdoor spaces.

PART 2

THE BUILDING

In climate-sensitive architecture, strategies are adopted to meet occupants' needs, taking into account local solar radiation, temperature, wind and other climatic conditions. Different strategies are required for the various seasons. In cold periods, for instance, a heating strategy has to be established which provides protection from the cold and takes advantage of available solar gains. In warm weather, on the other hand, a cooling strategy is required which gives shelter from the sun and, may in addition, make use of natural cooling techniques.

INTRODUCTION

In this chapter, the basic concepts needed for the formulation of heating and cooling strategies appropriate to the local climate are described together with the general principles of daylighting.

The Heating Strategy section covers:
- collection of the sun's heat through the building envelope;
- storage of the heat in the mass of the walls and floors;
- distribution of heat in the living spaces;
- retention of heat within the building.

The Cooling Strategy section covers:
- protection of the building from direct solar radiation;
- minimization of heat gains from external and internal sources;
- ventilation;
- natural cooling.

The Daylighting section covers:
- the general issues involved in daylighting;
- the energy implications of using artificial light;
- daylight and vision;
- definitions and units;
- the distribution of daylight in a room.

This chapter presents the basic principles of heating, cooling and daylighting requirements in building design. They are presented under separate headings for clarity but in practice will need to be integrated. It is intended that they should be used in the generation of the architectural form of the building, rather than be simply applied to a building which is already designed.

CHAPTER 4

THE BUILDING
HEATING STRATEGY

In cold weather, solar energy can make a positive contribution to the heating requirements of the building. For most parts of Europe it is appropriate to use the following four-part strategy:

- **Solar Collection:** solar energy is collected and converted into heat.
- **Heat Storage:** solar radiation has daily and seasonal cycles.Heat is collected during sunny periods and stored for future use.
- **Heat Distribution:** collected heat is distributed to the parts of the building which require heating.
- **Heat Conservation:** heat is retained in the building for as long as possible, as is any heat generated in an auxiliary heating system.

In those parts of Europe with suitable climates, exploitation of these principles can make use of an auxiliary heating system unnecessary.

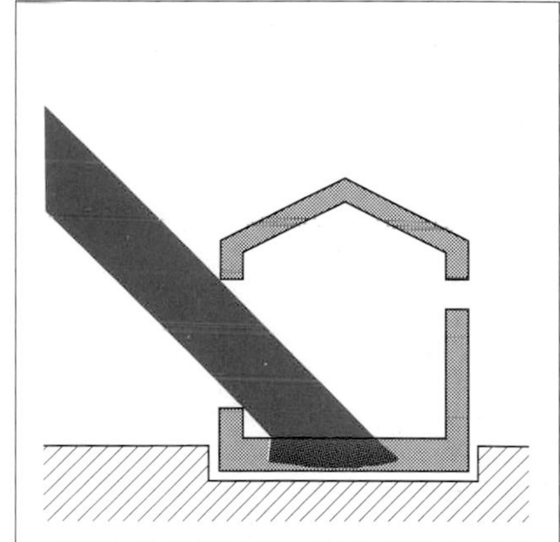

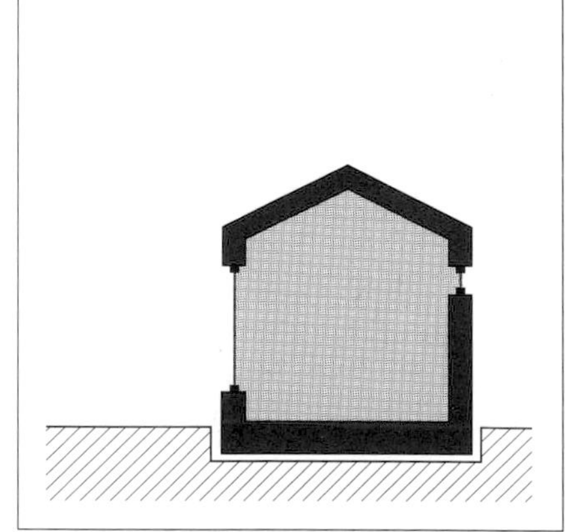

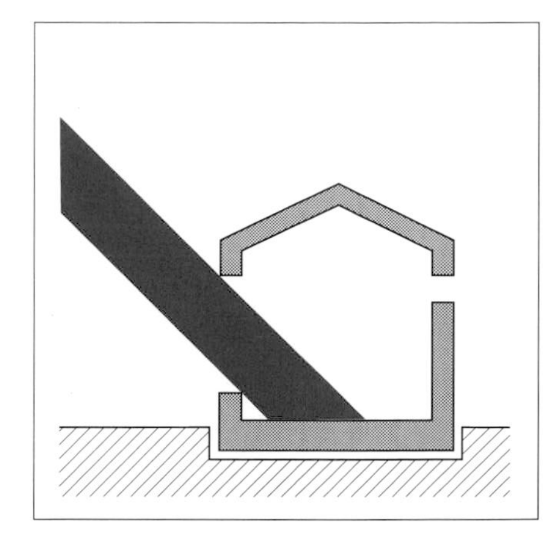

HEATING STRATEGY
SOLAR COLLECTION

The sun is essentially a gigantic fusion reactor which, in contrast to conventional, non-renewable energy sources, can supply an inexhaustible amount of heat. By collecting solar radiation, we can put this 'free' heat to good use, although there are costs involved in its collection.

TRANSPARENT ELEMENTS

When solar radiation strikes a transparent or translucent surface, part of it is reflected, part absorbed and the remainder directly transmitted.
The absorbed radiation is re-emitted to the inside or the outside of the building through convection or as long wave radiation. The proportion remaining within the building depends on the temperatures of the air and the translucent and adjacent surfaces and on the air speed on both sides of the element.
The total transmission through the element is the sum of the directly transmitted rays and the absorbed radiation re-emitted to the inside. The solar gains depend on what the element is made of, its area, the angle of incidence of the sun's rays and the available radiation which in turn depends on orientation, topography and shading.

The directly transmitted radiation will eventually reach another surface which may be translucent (another window) or opaque, for example a wall, floor, ceiling or piece of furniture. If it is translucent, part of the radiation will be reflected, part absorbed and the remainder transmitted, either to the outside of the building - if the obstacle is an external one - or to another room. If the obstacle is opaque, some of the radiation will be absorbed and the remainder reflected.

The solar gains through vertical glazing vary with orientation.
South-facing surfaces receive more solar radiation in winter and less in summer by comparaison with surfaces at other orientations. This, it will be noted, is in phase with the heating requirements of the building.
Throughout the year, the solar gains through west and south-west glazing are very similar to those through glazing facing east and south-east. In summer, windows facing west can give rise to overheating if they are not protected from the sun's rays, which are at a low angle of incidence.

The tilt of glazing also has an influence on solar gains. In summer, the gains through vertical glazing are lower than through glazing at other angles because the sun is high in the sky and the solar beam therefore has a high angle of incidence.
Glazing tilted at a low angle to the horizontal (e.g. 30 degrees) can cause overheating in summer despite the fact that it gives low gains in winter. In general, such glazing should be avoided unless it can be shaded efficiently when necessary. It can, however, be used in sunspaces or atria if they are partitioned off from the other occupied spaces in the building and have their own ventilation systems.

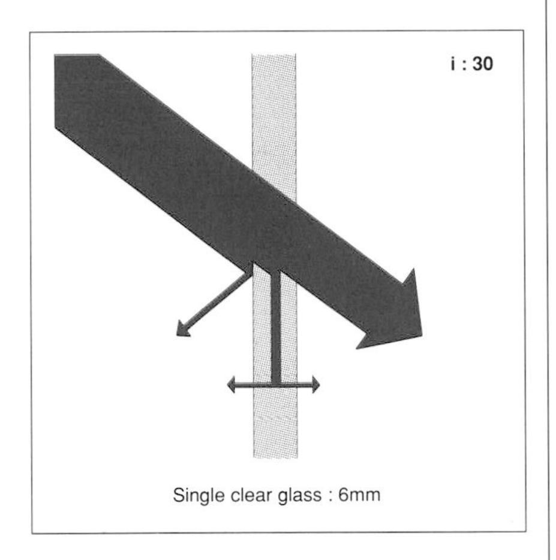

Single clear glass : 6mm

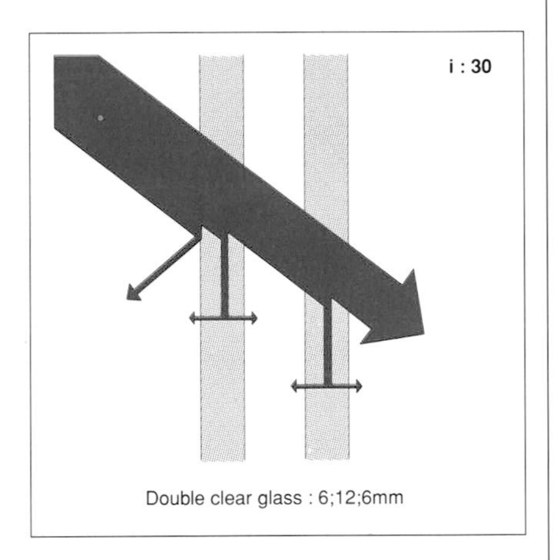

Double clear glass : 6;12;6mm

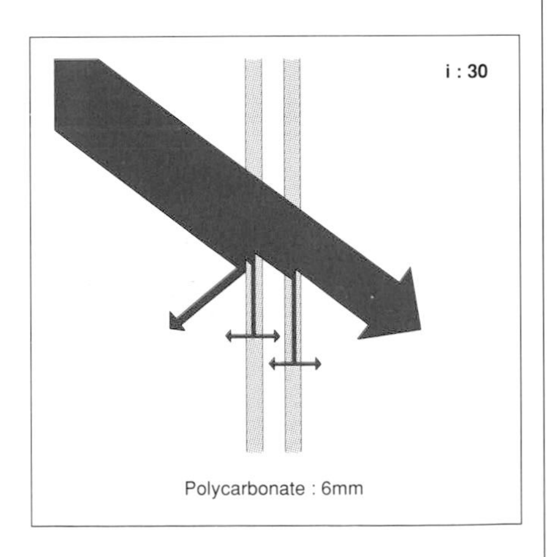

Polycarbonate : 6mm

OPAQUE ELEMENTS

When solar radiation strikes an external wall or other opaque surface, part of the energy is absorbed and transformed into heat and the remainder is reflected. There is no direct transmission.

Part of the absorbed energy is diffused towards the inside of the wall. The remainder is lost either by emission as infra-red long wave radiation to the sky or nearby surfaces or by convection to the outside air.
The amount of solar energy absorbed by the wall depends on the amount of incident radiation, the angle at which it strikes the wall, the absorbing capacity of the material the wall is made of and the condition of the wall surface. Dark, unpolished surfaces absorb more solar energy than do light-coloured polished ones.

The concept of collecting heat through walls is mainly applicable to warmer regions where there is a need for heating at night but thermal insulation is not necessary. In Northern Europe, more heat will be lost through an uninsulated south-facing wall than can ever be collected from the sun, measured either on a daily basis or over the whole heating season. In colder parts of Europe where outside walls therefore have to be insulated, the insulation prevents the diffusion of heat into the wall.

The process of collecting heat in a floor slab is similar. Where the floor is carpeted, the carpet serves as an insulating layer.

The diagrams opposite show the surface temperature variations of a range of south facing walls at six-hour intervals. A typical solid floor located behind south facing glazing is also shown. Internal temperature in all cases is a constant 20°C, and the curves represent clear sky conditions at 51°N latitude on 15 March with a 4m/s wind speed. The starting points of the curves show external temperatures.

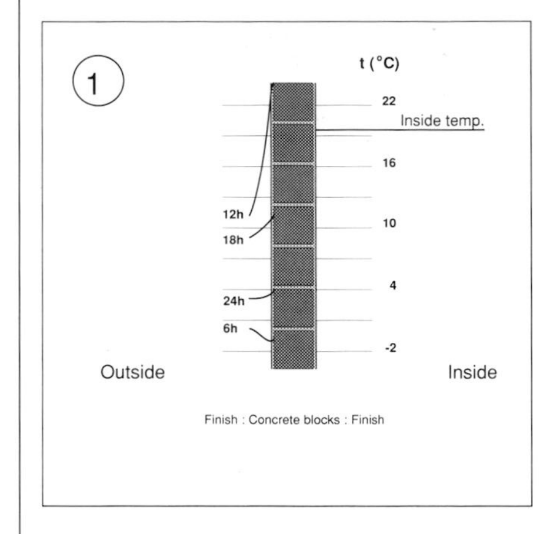

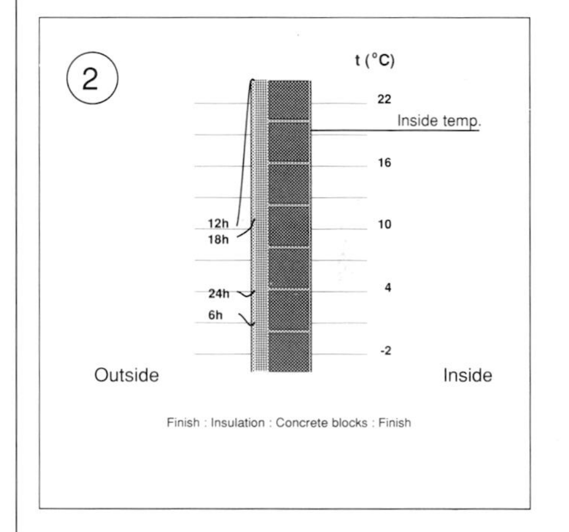

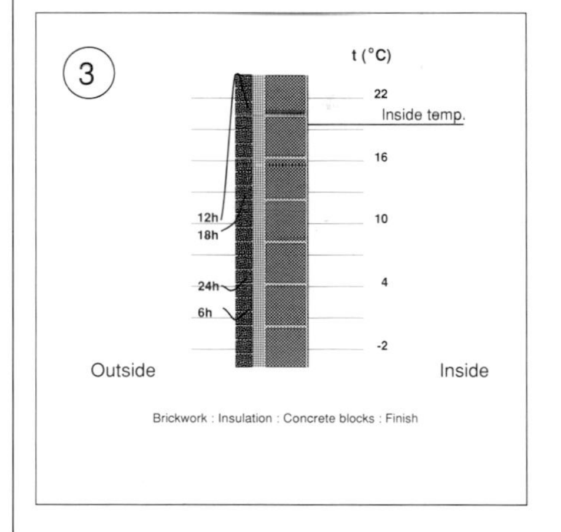

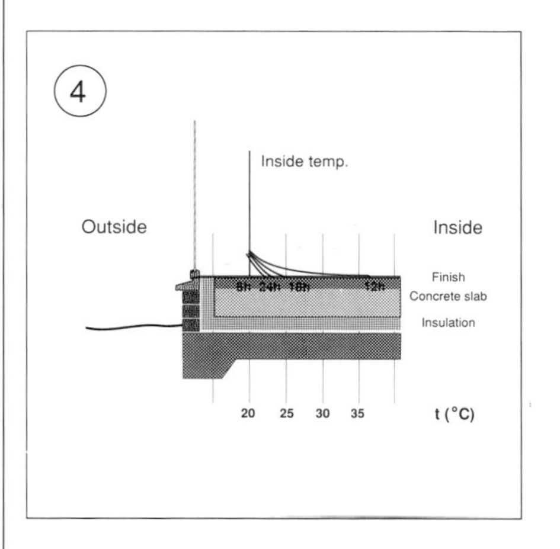

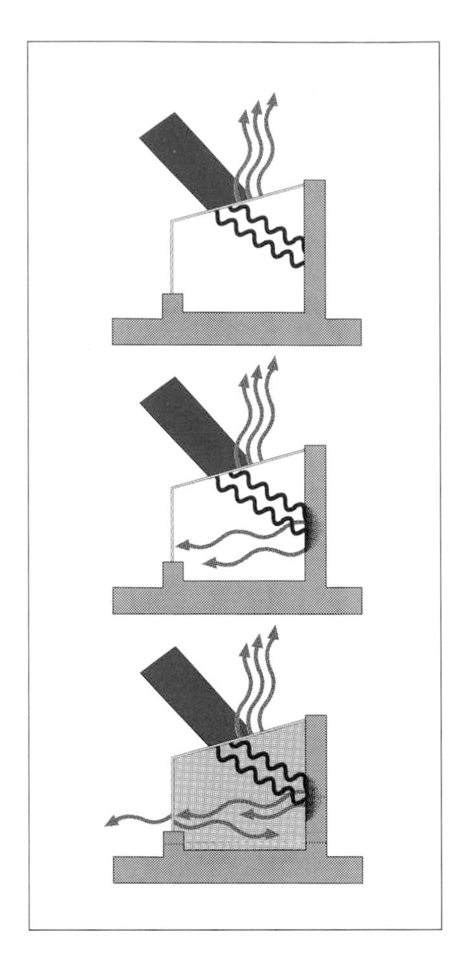

GREENHOUSE EFFECT

The greenhouse effect results from a three-stage process whereby shortwave solar energy is collected through glazing in the building envelope, absorbed by the opaque or solid elements in the building, and re-emitted as long wave radiation which is prevented by the glazing from leaving the building.
The efficiency of this collecting system is affected by its geometry, the characteristics of the glazing, (for example, the percentage of glazed area and the spectral transmission curve) and those of the solid elements struck by the solar radiation, such as solar absorptance and spectral thermal emission curve.

The solar radiation reaching the earth's surface has wave lengths in the range 0.25 microns to 4 microns (a micron is one thousandth of a millimetre), and is composed of ultraviolet (UV), infra-red (IR) and visible light. When the sun's rays strike glazing, much of the visible and short wave IR radiation is transmitted but most radiation with a wave length above 2.5 microns is blocked. The radiation which does pass through is absorbed by the walls, floor and other solid elements which heat up and then re-emit long wave IR radiation in all directions. When this strikes the glazing, part of it is reflected and the rest absorbed. The absorbed energy is then re-emitted on both sides of the glazing. As a result of this process, part of the incoming radiation is trapped inside, creating a rise in temperature, which is known as the greenhouse effect.

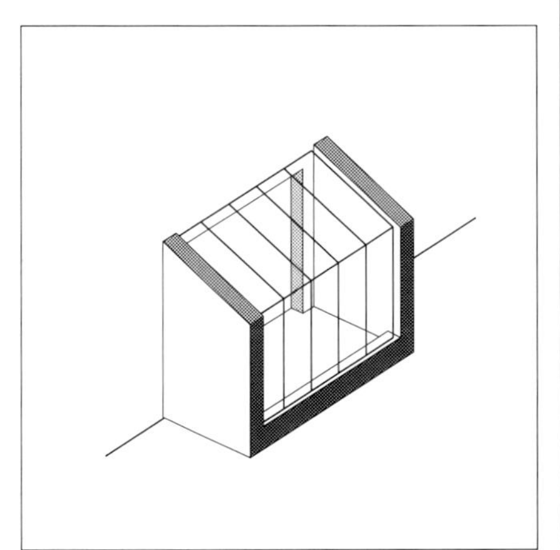

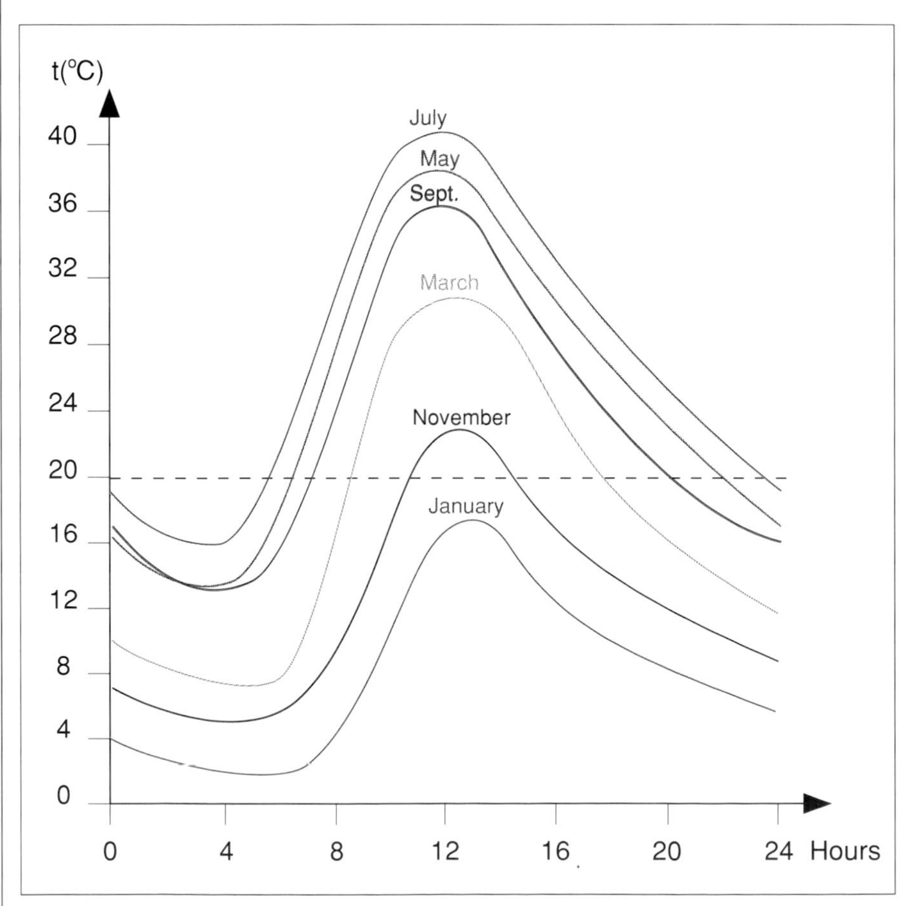

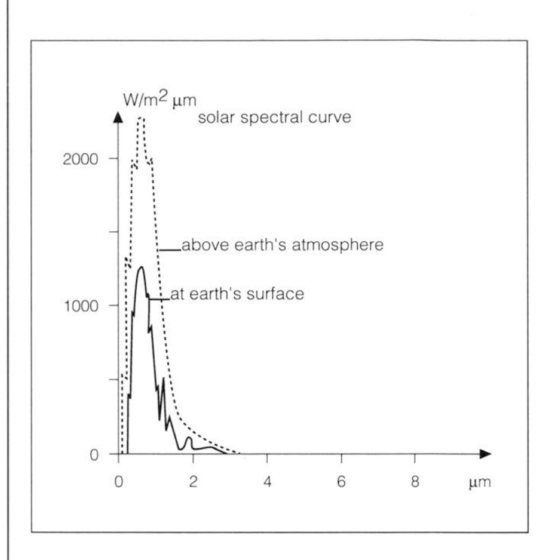

With double glazing, the total transmission is lower than with single glazing. The emission from the walls behind the glazing will therefore be reduced, since less radiation strikes them. With a double layer of low-emissivity glass, glass with a special coating which gives it a low emissivity to long wave (thermal) radiation, the transmission will be lower still but a greater proportion of the thermal energy will be retained in the building.The graphs opposite show temperatures under clear sky conditions in an unheated, single-glazed attached sunspace facing due south at latitude 51° N. The sunspace has a $10m^2$ floor area, a volume of $25m^2$ and is attached to a building in which the internal temperature is 18°C. There is double glazing between the sunspace and the building. During autumn and spring the sunspace can supply heat to the building, but in summer, shading is required to avoid overheating.

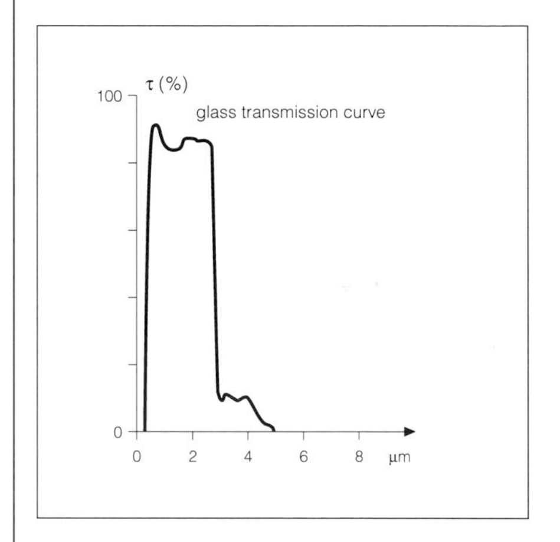

It should be noted that any auxiliary heating of the sunspace during the heating season will destroy the benefits of solar heating and will change the sunspace from an energy saving feature into an energy using one. It is necessary to separate the sunspace from the rest of the building with a solid or glazed wall and doors in the wall should remain closed whenever possible.

The same sunspace during March with different orientations.

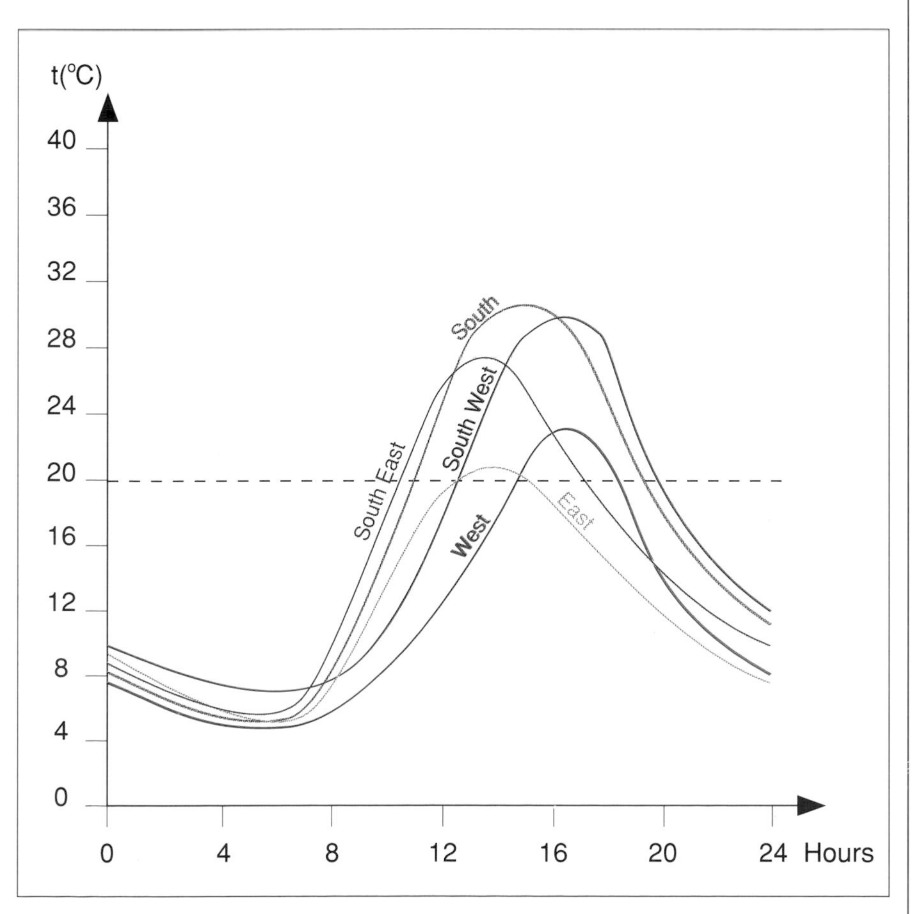

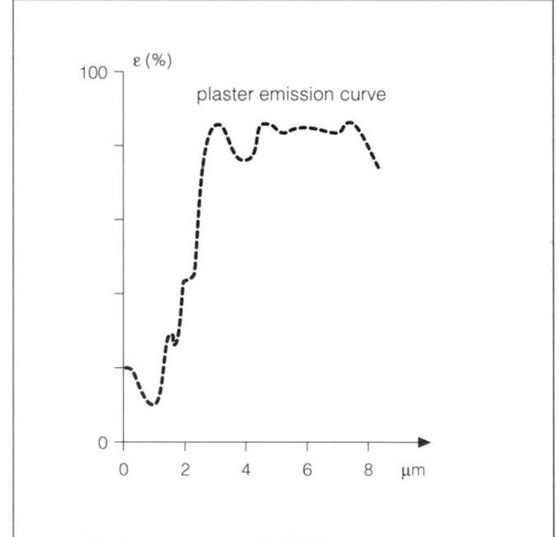

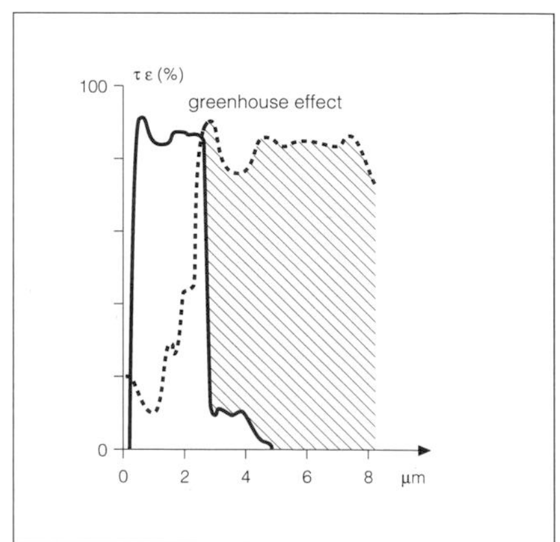

HEATING STRATEGY
HEAT STORAGE

Because solar energy is emitted by the sun in diurnal and annual cycles, its availability is often not in phase with the heating requirements of the building. The aim of storage is to retain the heat collected which is surplus to current needs in order to use it later, when required.

DIRECT STORAGE

When solar radiation strikes a material - either directly or after transmission through glass - part of it is absorbed, transformed into heat and stored in the mass of the material. The material heats up progressively by conduction as the heat diffuses through it.

Heat penetration is quickest in materials with a high thermal diffusion coefficient. This increases with increasing conductivity. Thermal diffusion in the material prevents the surface temperature from rising rapidly when radiation falls on it and causes the temperature of the entire mass to increase. Materials with high heat storage capacity such as concrete, brick and water heat up and cool down relatively slowly. Thermal insulating materials such as glass fibre and foam, usually because of their open or cellular structure, form poor heat stores and diffuse heat very badly. The insulating layer of a wall minimises heat exchange between the solid layers adjacent to it.

The diagrams opposite show temperature variations at six hour intervals and the heat accumulated per square metre at 2pm. All of the walls and the glazing adjacent to the floor are oriented due south under clear sky conditions with a wind speed of 4 metres per second at latitude 51°N on 15 March. The internal temperature is a constant 20°C and the external temperatures at 6 hour intervals are also shown.

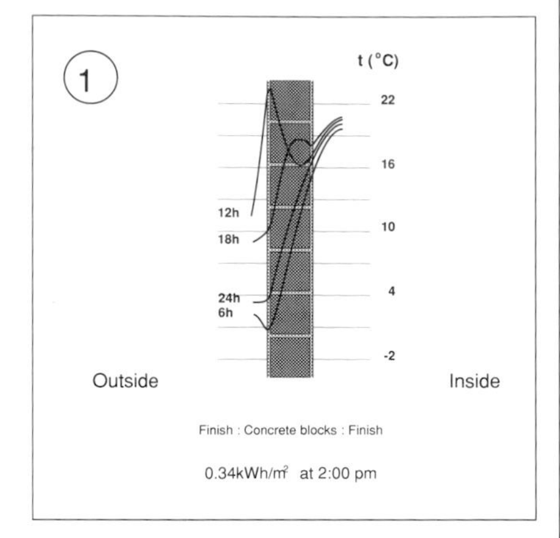

Finish : Concrete blocks : Finish

0.34kWh/m² at 2:00 pm

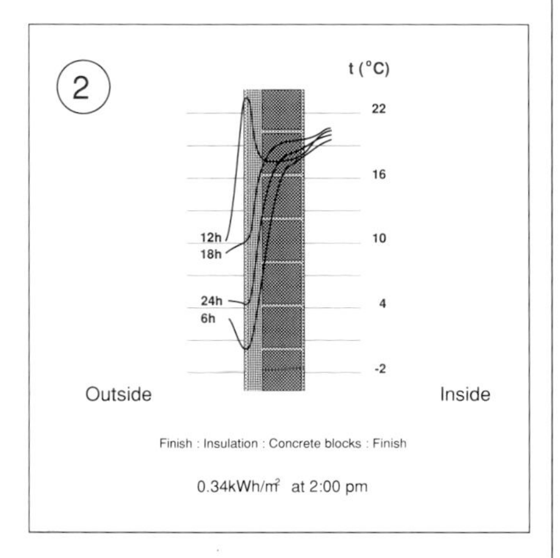

Finish : Insulation : Concrete blocks : Finish

0.34kWh/m² at 2:00 pm

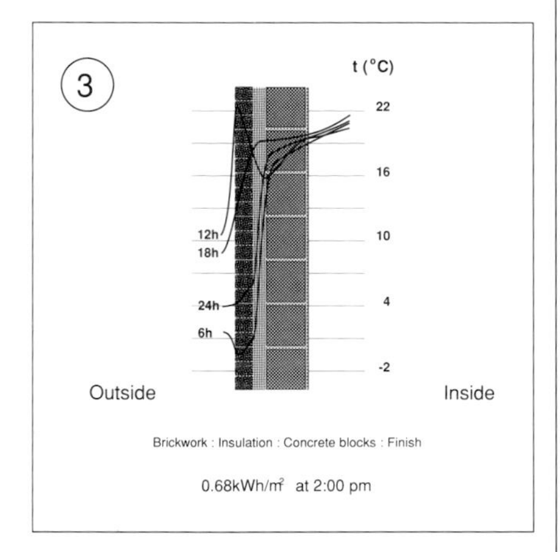

Brickwork : Insulation : Concrete blocks : Finish

0.68kWh/m² at 2:00 pm

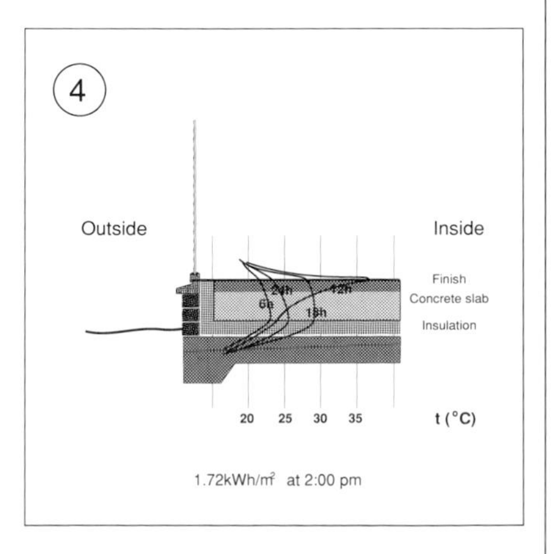

1.72kWh/m² at 2:00 pm

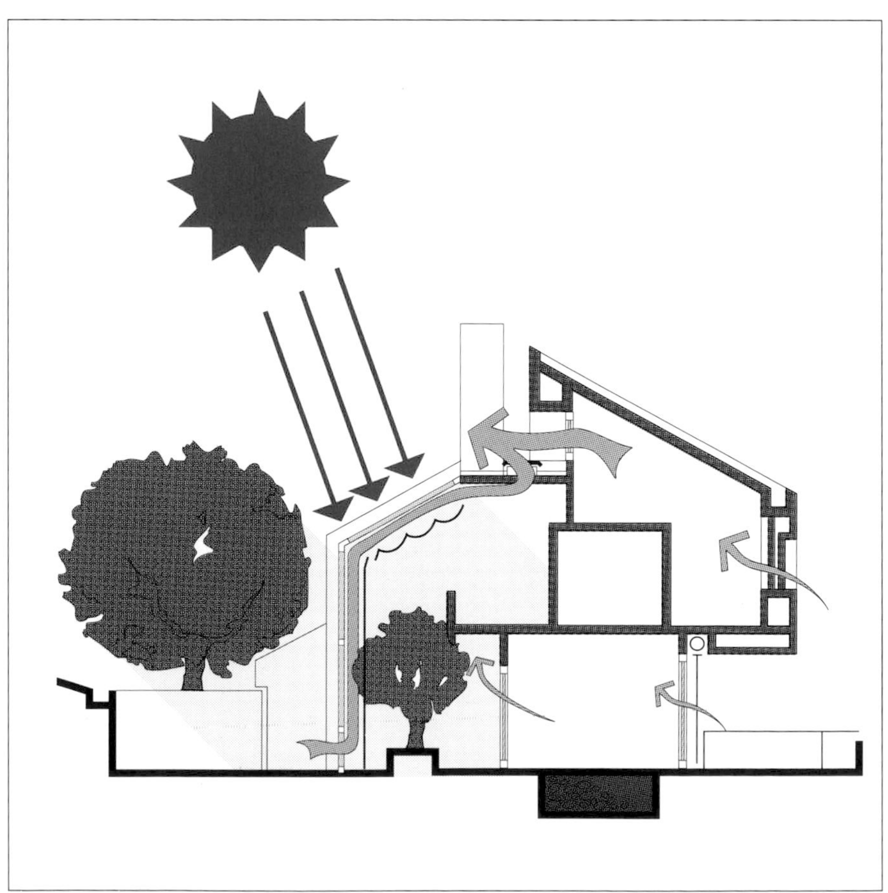

Daily variation of the heat quantity stored through solar radiation during different months, under clear sky conditions, in a floor similar to the one shown on the previous page is shown to the right.

This floor is immediately under a south oriented window. Inside temperature is kept constant at 20°C, while temperature under the floor is 13°C in March, 16°C in June, 17°C in September and 14°C in December.

The floor stores most heat in March and September because solar radiation is less in December and because the angle of incidence of the solar radiation on the window is greater in June causing more of the floor to be in shadow.

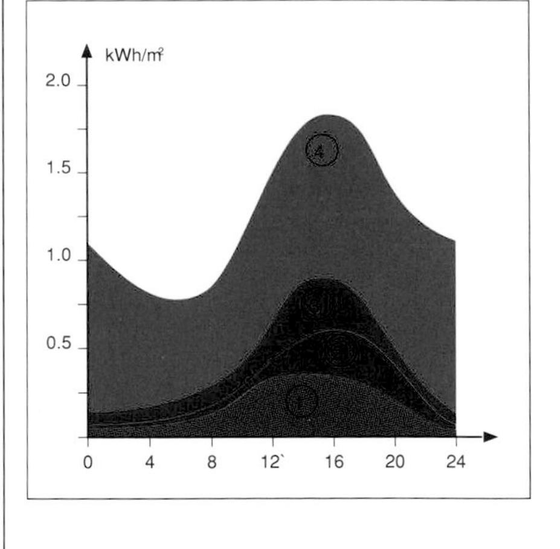

Storage of solar gains for the month of March, under clear sky conditions, in the same floor type but with different colours; typical factors are 0.9 for black or dark blue; 0.7 for red or brown; 0.5 for dark grey or green; 0.2 for white. The inside temperature is kept constant at 20°C and the temperature under the floor is 13°C.
The colour of the surface material greatly influences the solar collection and thus the storage characteristics.

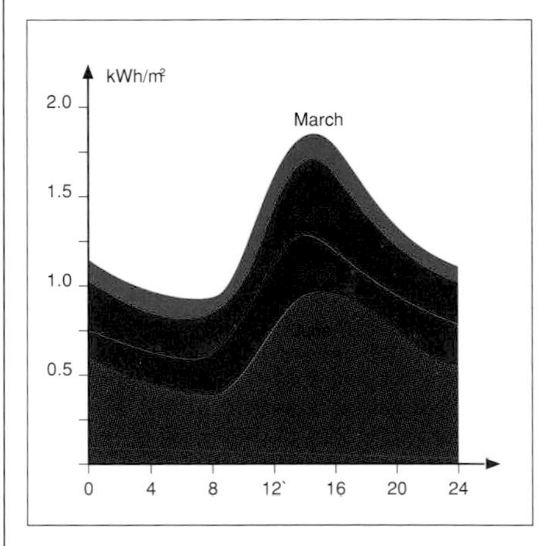

Storage of solar gains for the month of March, under clear sky conditions, in the same floor type, but with different thickness of concrete slab. The inside temperature is kept constant at 20°C while the temperature under the floor is 13°C.

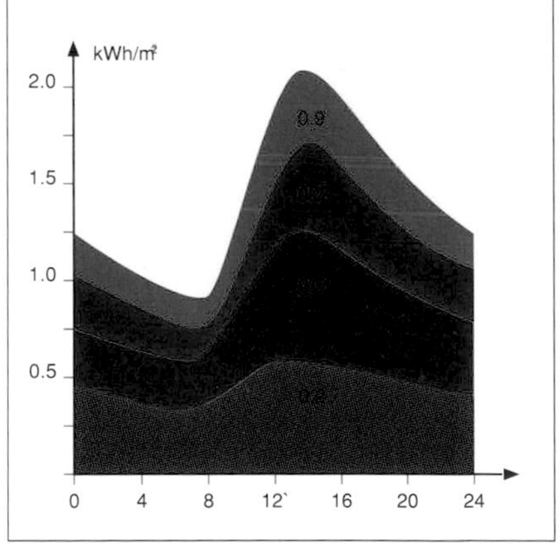

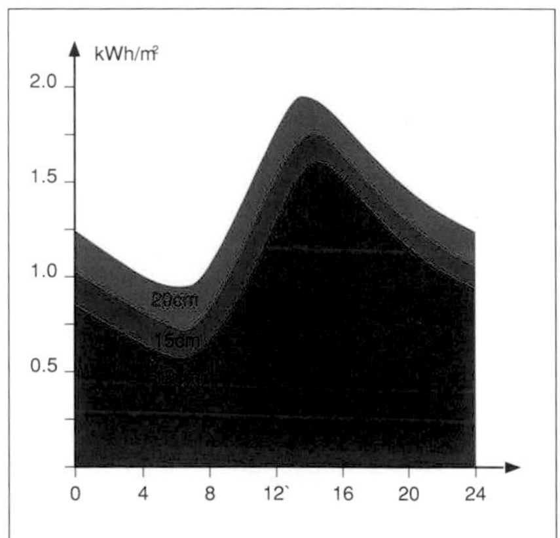

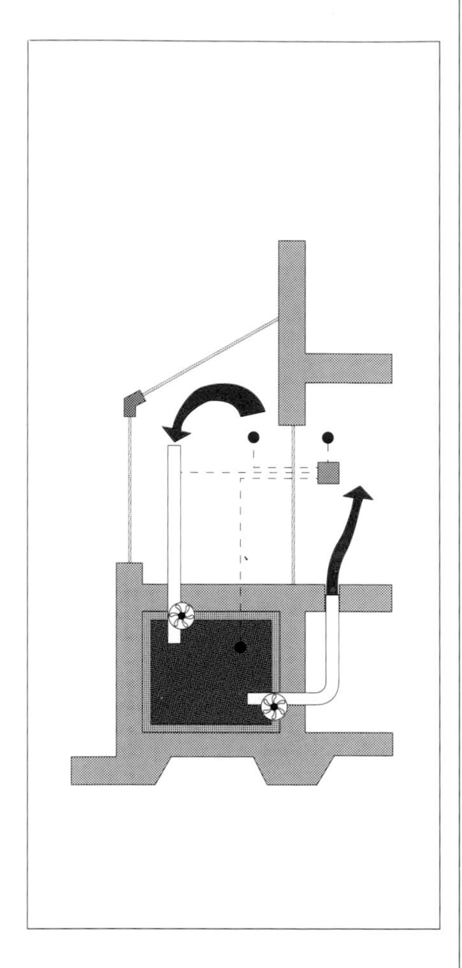

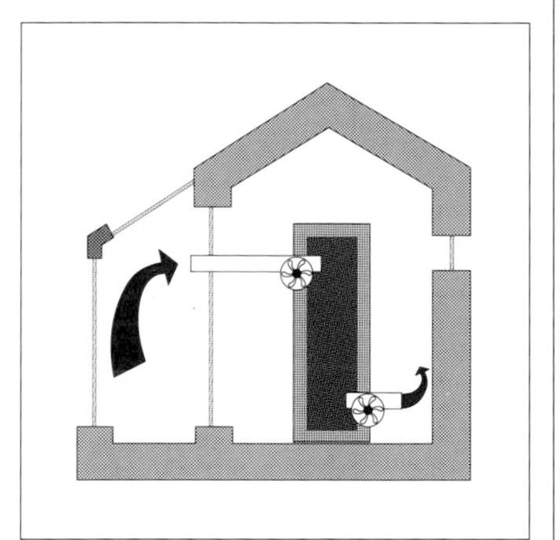

INDIRECT STORAGE

When a building component is heated up by absorption of heat radiated from other, warmer, components (such as walls and floors) or by convection from the surrounding air, this is known as indirect storage.
Indirect storage through radiation is influenced by the temperature difference between the components, their location and emissivity. Unlike visible radiation, the emission of infra-red radiation is not affected by the colour of the surface. It is, however, influenced by surface condition, as rough or dirty surfaces inhibit emission.
Indirect storage through convection is influenced by the temperature difference between the air and the component, by the speed of the air and the roughness of the component surface. Rough surfaces have a greater area and therefore assist convective heat transfer.
Heat exchange between masses at different temperatures is spontaneous because it obeys the second law of thermodynamics concerning thermal equilibrium between masses.

Sometimes, it may be desirable to make use of a remote storage system to accumulate heat transferred from a sunspace for instance by fans and air ducts. Such systems usually aim to diffuse heat to the building in a controlled way, as required. They will only be effective if the store is sized correctly and insulated well so that a significant increase in store temperature is achieved. In costing such items, it should be noted that the fans will consume power.
To achieve thermal efficiency, it is better to locate a mass store within a building so that any heat loss from the store can be collected by the building. This can, of course, take up space which might be better used for some other purpose.

STORAGE MATERIALS

The ability of a material to store heat is usually expressed in terms of its specific heat capacity - the heat stored in a unit volume of material per degree of temperature rise. Materials with a good thermal conductivity accumulate heat relatively quickly. Dense materials such as natural stone, concrete and brick are traditionally chosen for the parts of a building where good heat storage is required. The storage capacity of masonry ranges from 0.204 kWh/cubic metre for cellular concrete to 0.784 kWh/cubic metre for heavyweight concrete.

In a lightweight building or for parts of any building where increased storage is required without the use of massive materials, substances with a higher storage capacity can be used. Examples are water (which has a storage capacity of 1.157 kWh/cubic metre at 20 degrees C), other liquids, and materials which change phase.

With traditional materials such as concrete and brick, the storage process makes use of sensible heat, that is, heat which can be measured as it results in an increase in temperature. Phase-change materials, on the other hand, make use of the latent heat of fusion, that is the heat required to change the state of the material from a solid to a liquid without a change in temperature. In the construction field, it is usual to choose a material which changes phase at a temperature somewhere in the range from 2 degrees C to 50 degrees C. Very large quantities of heat (typically 38 to 105 kWh/cubic metre) can be stored when the phase change occurs. Therefore a much smaller volume is required for a phase-change store than for a conventional store. In addition, undesired thermal losses are avoided because the temperature increase during the storage period is very low.

New materials are currently being developed which change their molecular structure without changing state. This transformation also uses latent heat and can therefore be used to store heat. The advantage of the new materials over phase-change materials is that they remain solid. Currently, however, they are expensive and can only be used over a limited number of charge-discharge cycles. Their reliability over longer periods of time has to be improved before their use can become widespread.

The efficiency of phase-change and related stores depends on the durability over time not only of the storage material but also of its surrounding envelope. The latter has to withstand considerable volume changes.

When dealing with liquids, special care must be paid to the possibility of freezing and problems related to thermal expansion.

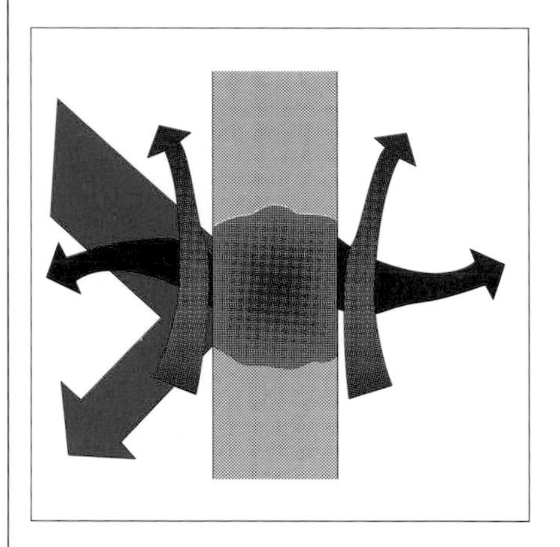

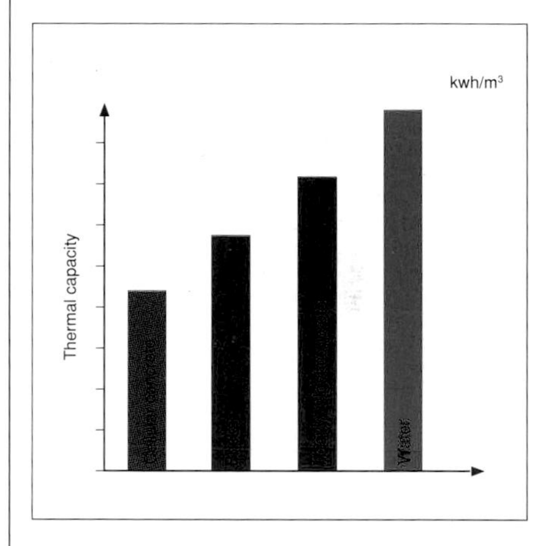

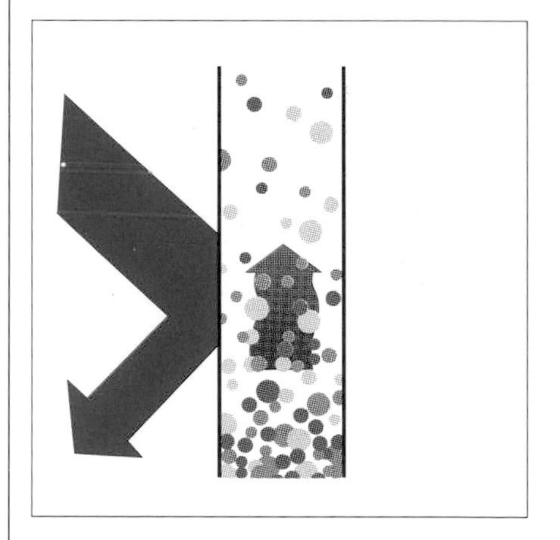

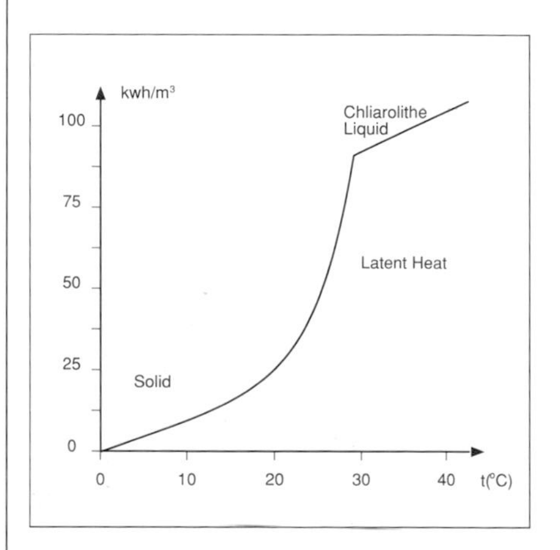

HEATING STRATEGY
HEAT DISTRIBUTION

Ideally a passive solar building should provide heat directly to the areas where it will be used, thus requiring no distribution system. However, this is not always possible as in the case of north facing rooms where stored heat must be distributed at the appropriate time. This can be achieved by natural or mechanical means.

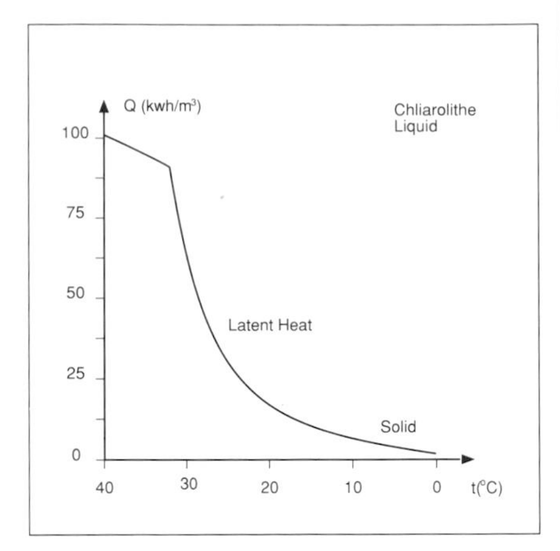

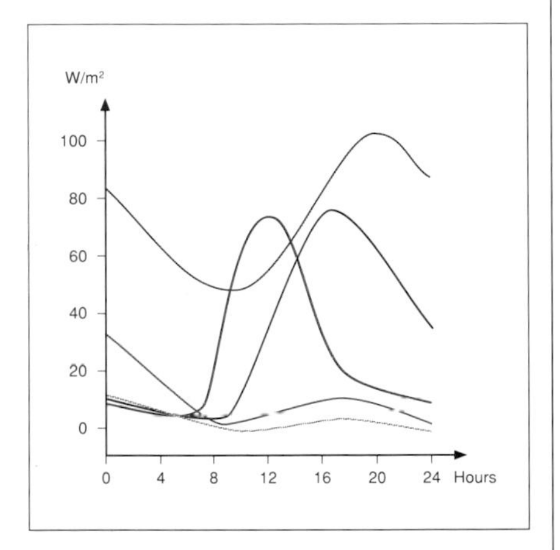

NATURAL DISTRIBUTION

In natural distribution, the stored heat is transmitted by convection and radiation. Convection occurs when the surface temperature of the storage material is above ambient. Long wave infra red radiation emission takes place when the surface temperature of the storage material is higher than the surface temperature of neighbouring objects.
When energy has accumulated in the wall of a building, the diffusion of heat will be almost immediate on the side exposed to the radiation. On the opposite side, however, there will be a time delay before heat is released. The delay or lag, which is defined as the time between the moment when the irradiated face of the wall reaches its maximum temperature and the moment when the opposite face reaches its maximum temperature, is influenced by the thermal inertia of the wall. The extent of the delay depends on the dimensions and physical properties of the wall. If there is thermal insulation in a wall, each part of the wall can be regarded as having its own thermal inertia.
When the inside face of the wall reaches an appropriate temperature, the air of the space beyond will be heated by convection and surfaces near the wall will be heated by radiation. As a result, heat will have been distributed from the wall to spaces which cannot benefit from direct solar gain. This delayed heat transfer process can help to maintain comfortable temperatures for a significant length of time after the solar radiation has ceased. As a result, the building's heating requirements will be reduced.
A wall of high thermal inertia will allow heat stored during the day to be released at night. An uninsulated wall of this type is particularly useful in warm climates where living spaces only need to be heated at night.

Heat stores containing phase-change materials release a great deal of stored heat without much change of temperature as the material passes from the liquid to the solid state. The stores then behave as described above, releasing sensible heat to their surroundings in proportion to the temperature difference.

The diagram opposite shows the quantity of heat diffused per second per square metre of the walls and floor described on page 54.
The internal temperature is a constant 20°C and the temperature below the floor is 13°C.

Comparisons may be made between walls. The floor behaves differently as it absorbs and re-emits heat on the same side.

THERMOCIRCULATION

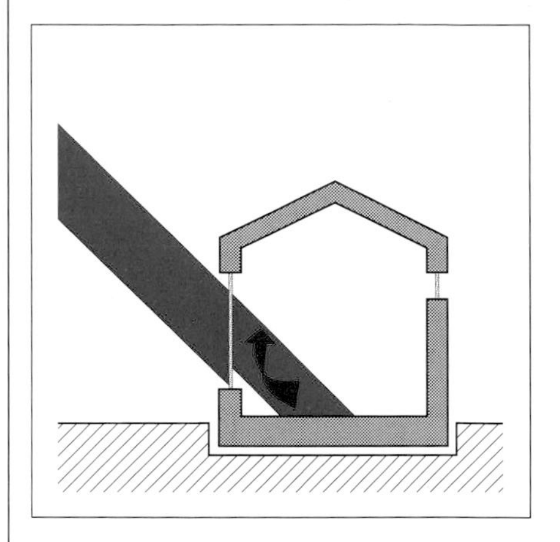

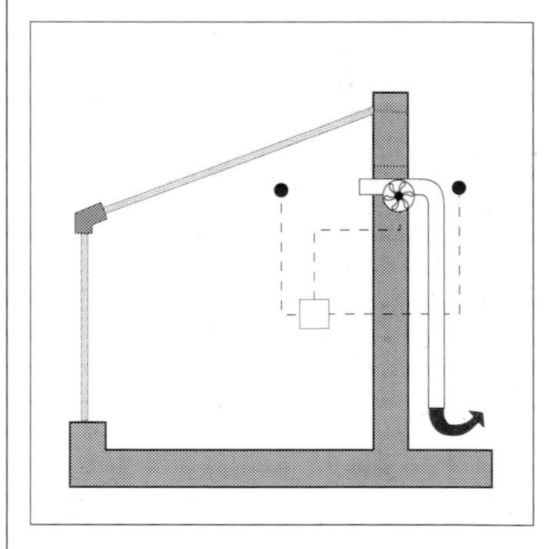

Another form of natural circulation is achieved when air heats up. When this happens, the air's density decreases and it tends to move upwards. Judicious organization of spaces in a building allows this phenomenon to be used to distribute heat generated by direct solar gain in one zone to another, cooler zone.
When solar radiation heats up an inside wall, the stored heat is released and heats up the air in the room by convection. As this air becomes warmer, it rises and its place is taken by cooler air. When the warm air reaches a place unheated by direct solar radiation, it becomes cooler. Meanwhile, the cooler air in contact with the irradiated surface is heated up. An air circulation loop is thus set up between the zone which is directly heated by solar radiation and the non-irradiated zones, as long as the organization of spaces permits this. This air movement - which can be controlled by the opening and closing of doors and interior windows - is called thermocirculation.
In cloudy periods or at night, air movement may need to be prevented by closing off the zones or by using movable insulation - otherwise a reverse loop could be established, creating unwanted cooling.

MECHANICAL DISTRIBUTION

Heat can also be distributed by using mechanical devices such as fans or pumps. Immediate distribution can be produced by using a fan to force hot air from spaces heated by direct solar gain to cooler spaces. Delayed distribution can be achieved when there is a storage mass in the building. A fan is used to drive heat from a solid store to zones requiring heat. In the case of liquid stores, a pump is used.

The combination of passive solar and conventional warm air heating systems can be useful, allowing excess warm air to be moved to cooler areas of the building.

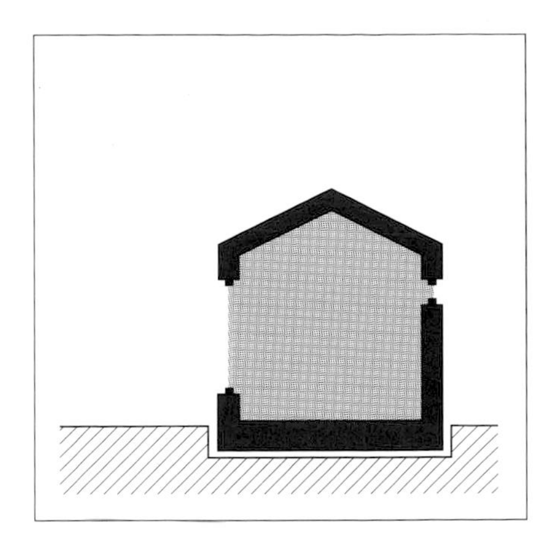

HEATING STRATEGY
HEAT CONSERVATION

The building envelope can lose heat by transmission through thermal conduction, convection and radiation processes or by ventilation or unwanted infiltration of air. In cooler weather, these losses need to be minimized so that heat originating from solar gains and the auxiliary heating system are retained in the building as long as possible.

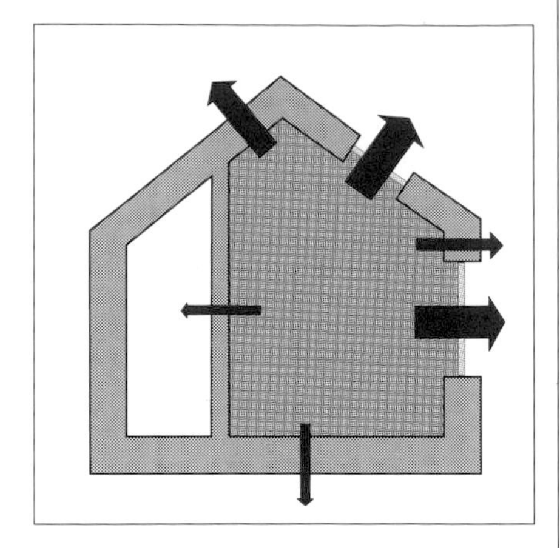

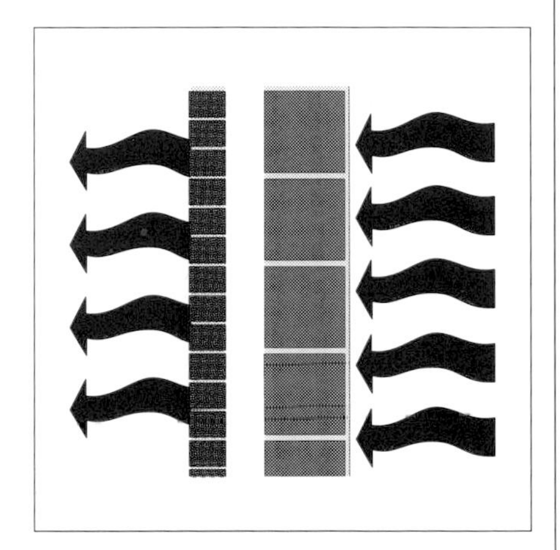

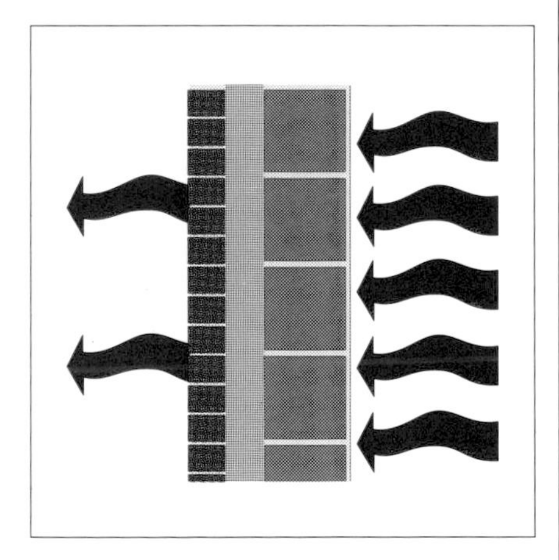

REDUCTION OF TRANSMISSION LOSSES

Transmission losses are stated in terms of heat flow through the envelope, that is the quantity of energy passing through the envelope per unit of time, and are usually given in Watts (Joules per second). They depend mainly on the temperature difference between the inside and outside face of the envelope and the thermal resistance of the material (or combination of materials) of which the envelope is made.
These losses take place through conduction, convection and radiation and there are a number of ways of reducing them. Some examples are described here.
The most common method is to prevent heat conduction by adding thermal insulation to the envelope to increase its thermal resistance.
Another way of reducing transmission losses is to design the building more compactly to reduce the amount of surface through which heat can be transmitted.
Additionally, the inside temperature of the building can where practical be lowered in order to reduce the temperature difference between the building interior and the outside air and thus reduce heat transmission.
A further method is to add barriers to radiative heat flow by, for example, placing aluminium foil behind radiators and using low-emissivity glazing. These cut down on exchange of heat between two components by reflecting the long wave IR radiation back into the room.

The thermal resistance (R) of a wall is equal to the sum of the thermal resistances of its components plus coefficients for the wall-to-air films at the two faces of the wall. A higher R-value means that there is a greater resistance to heat transmission through the envelope.
The thermal resistance of a layer of material depends on the thermal conductivity of the material and the thickness of the layer.
The thermal conductivity is the quantity of heat passing at steady flow through a one metre thickness of a square metre of the material in one second when the temperature difference between the faces is 1K. It is given in Watts per K. By convention temperature intervals are expressed as K (Kelvin), whereas actual temperatures are expressed as °C (degrees centigrade)
The factors which influence thermal conductivity most are density, water content, pore size and type of material around the pores.
The film coefficients contribute to thermal resistance because heat has to change its mode of propagation from conduction to convection and radiation or vice versa at each surface of the wall. They depend on various parameters such as roughness of surface, temperature difference between walls or between the wall and the air, air speed, direction of heat flow and radiation characteristics of the surfaces under consideration, and adjacent surfaces.

OPAQUE ELEMENTS

To reduce transmission losses as much as possible, the walls, roofs and other opaque elements of the building must be provided with permanent thermal insulation. This improves the thermal properties of these elements and, as a result, the elements are maintained at a higher temperature than would be the case in an uninsulated building. This improves the comfort level.

The insulation serves as a barrier to heat flow by conduction. Materials chosen for this purpose must, therefore, have a low thermal conductivity. The best common insulation material is air if it remains dry and still: it is important that there is no convection.

Porous materials with many small air pockets also make good insulants. Again, it is important for the materials to remain dry. When water fills the cavities their conductivity increases rapidly. (The thermal conductivity of air is 0.026 W per metre K; the thermal conductivity of water is 0.58 W per metre K.)

Insulation can be placed on the outside face of the wall, on the inside, or within the wall without altering the overall insulation properties of the wall. However, the wall's thermal inertia and the risk of condensation inside the building are affected by the position of the insulation. External insulation increases the useful thermal inertia within the wall and substantially reduces the risk of condensation in the building. It also reduces the possibility of problems from thermal bridges. These are low resistance paths connecting two surfaces where insulation is deficient or non-existent.

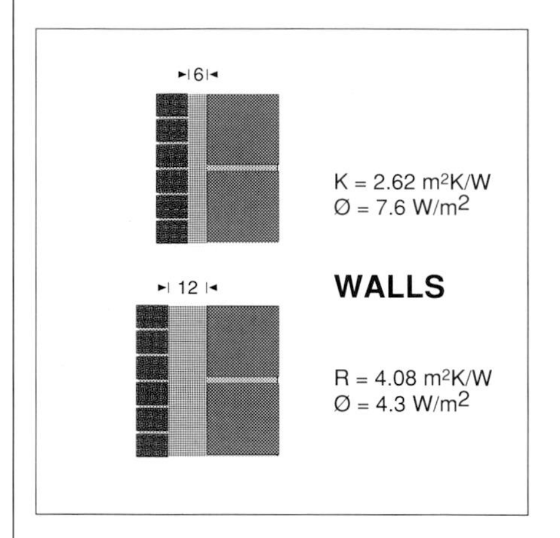

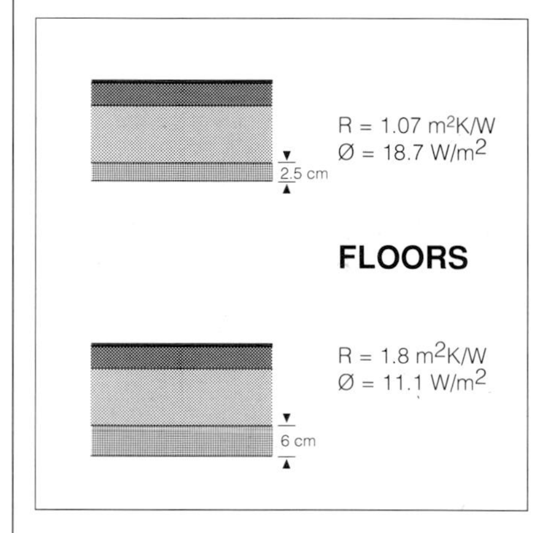

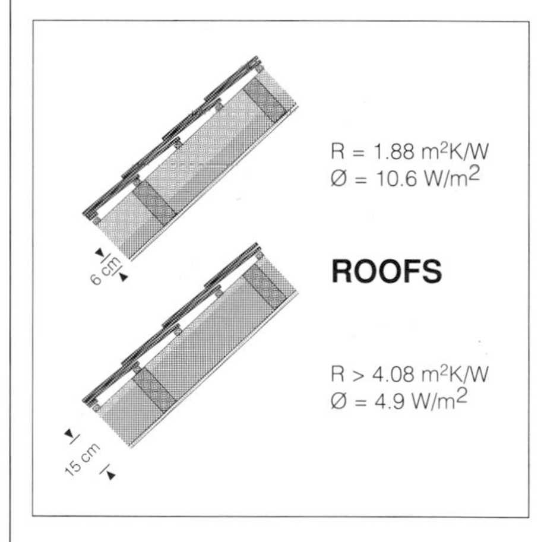

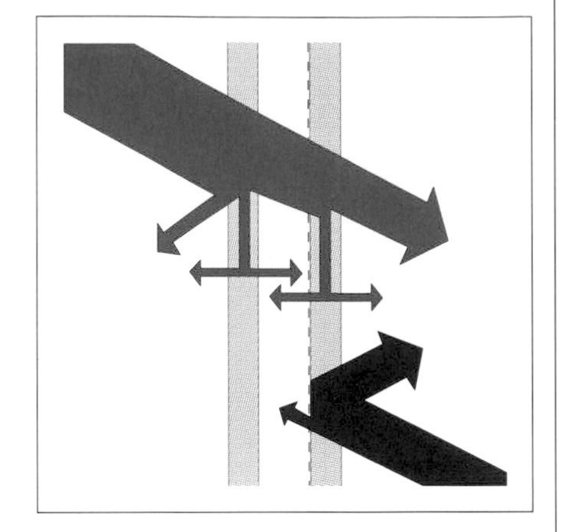

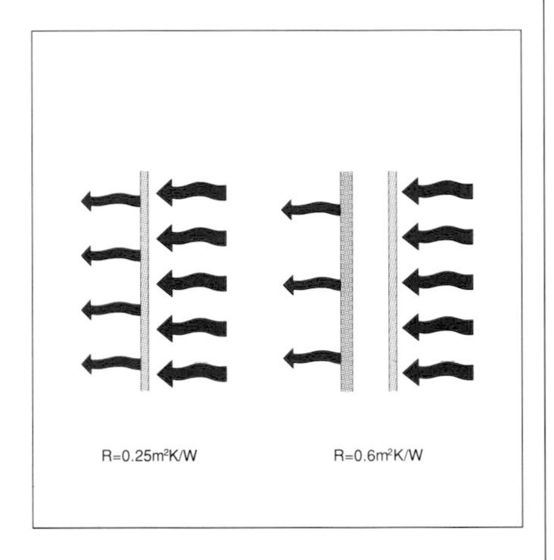

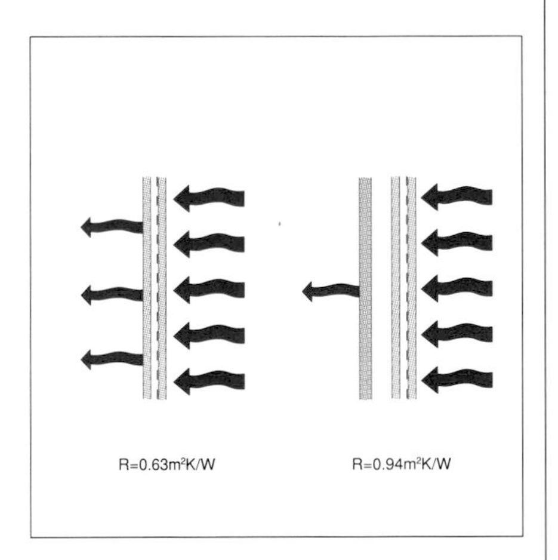

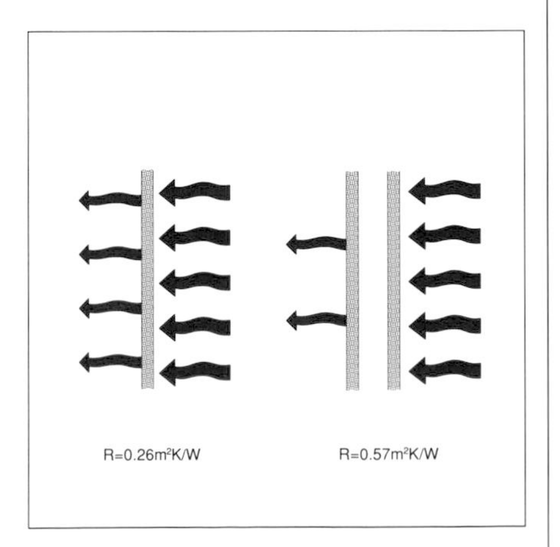

TRANSPARENT ELEMENTS

Transparent elements such as glazing can provide an easy path for heat to flow out of the building because of the relatively poor thermal properties of the transparent materials themselves and of their frames.

The transmission properties of the element can be reduced to a half or a third if it is constructed from two or three layers of glazing (instead of one) and the gap(s) filled with dry air or a special low conductivity gas.
Special treatment of the glass itself can also reduce thermal losses. For instance, the inside sheet of a double glazed unit can be coated on the side facing the cavity with a film of metal oxide having low emissivity properties. This film hinders the long wave IR radiation from leaving the building while allowing solar radiation to enter in a normal fashion. The metal oxide serves as a barrier which reflects the IR radiation into the room. The thermal transmission of this 'low-e' glazing is about 30% of that of ordinary single glazing. It almost equals the performance of triple glazing.

Increased thermal resistance of a transparent element can also be achieved by use of curtains, shutters or other movable insulation. The use of transparent insulation, which is now becoming commercially available, may be preferable because it reduces the thermal transmission losses from the building while allowing the light to continue to be transmitted.

In choosing insulation for windows, it is better to select external devices rather than those which fit inside the building. This is because over and above their insulation properties, such units stop heat loss at the surface of the glazing by convection or radiation to the sky. They can also be used to provide protection from the sun and guard against overheating.

Glass manufacters are currently developing evacuated double-glazed units which can reduce heat transmission to one tenth of that of a single pane of glass

REDUCTION OF INFILTRATION LOSSES

The renewal of air in buildings is necessary to eliminate stale air, smoke and odours, to control pollutants, and to maintain comfortable levels of oxygen and humidity. Minimum hourly air replacement rates are usually set at around 20 cubic metres per occupant. The actual figure depends on region and type of occupancy and is generally stated in terms of the number of air changes per hour (ACH), i.e. the volume of fresh air introduced, expressed as a multiple or a fraction of the total volume of building space under consideration. Typical domestic levels are 0.5 to 2 ACH, but these may need to be significantly higher in bars or restaurants for example.

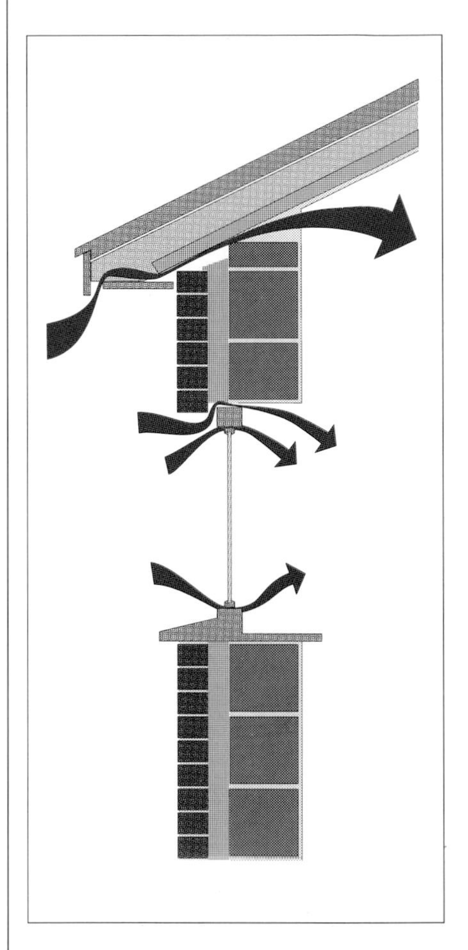

Infiltration of air into a building results from holes (such as open windows, chimneys and ventilation ducts) in the envelope, gaps between building components, joints around the movable parts of doors and windows and penetration of air through building components under pressure from wind. The level of infiltration depends on wind speed and pressure and temperature differences between the outside and inside of the building. It is difficult to estimate what the rate of infiltration of air into a building will be but it is possible to measure it once the building is constructed.

It is not necessary to cut out air infiltration entirely. The aim should be to minimize it so that unnecessary ventilation is avoided and replacement of air can be controlled easily. Care should be taken over any aspect of the design which could have an effect on wind penetration. For instance, thought should be given to topography, building shape, planting of wind shelter, etc. In construction, workmanship should be good and attention paid to details such as joints and closing systems, etc., around the critical parts. Infiltration through windows can be reduced by making joints air-tight or by using insulation devices such as shutters. Once efforts have been made to reduce infiltration, further conservation of heat can be achieved if occupants are made aware of losses caused by unnecessary ventilation.

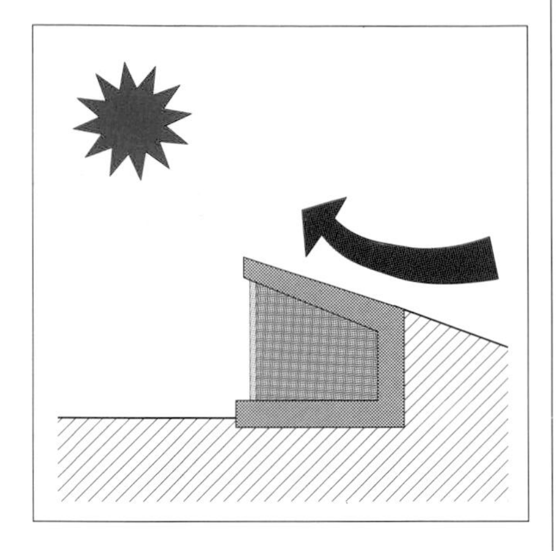

USE OF TEMPERATURE DIFFERENCES

When the temperature difference between the inside and outside of the building envelope increases, the transmission and infiltration heat losses through the envelope rise correspondingly. Because the temperature of the outer surface rises when it is exposed to solar radiation and decreases when it is subjected to strong winds, the designer should reduce as far as possible the area of building envelope facing north and the amount facing the prevailing winds.

Earth-sheltering a building will also reduce transmission and infiltration heat losses, especially in areas where there are strong winds and low ambient temperatures. This is because ground temperatures remain fairly constant throughout the year and the soil gives an additional thermal resistance to the building envelope. However, this will not be of much benefit if insulation standards are already very high.

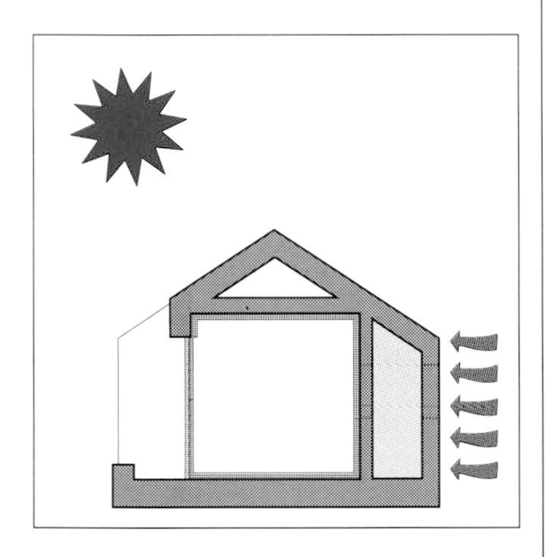

It is quite possible to create a building where each room is comfortable with respect to the type of activity carried out there. To achieve such a design, the hours of the day during which the room is used, its solar gains and its internal gains from artificial lighting, other equipment and occupants must all be borne in mind.

In organizing the building spaces, it is sensible to make use of the concept of 'thermal zoning' to create a rational distribution of heat and to reduce thermal losses. One way of achieving this is to face the rooms with a high energy demand south and to put the other spaces on the north side of the building. Alternatively, the 'hot core' of the building can be surrounded by concentric circles of rooms with decreasing temperature requirements.

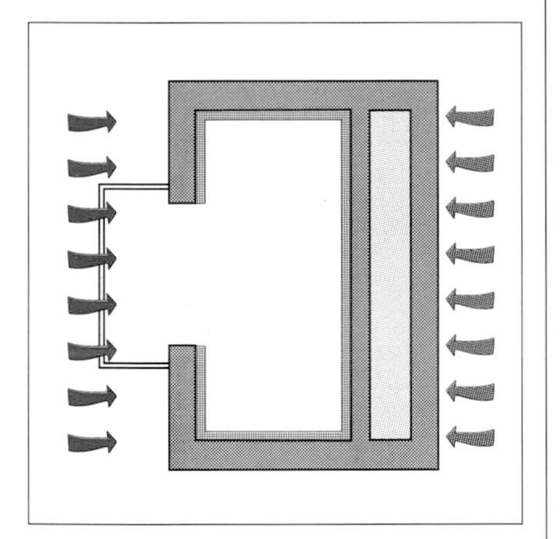

These principles are used when unheated spaces such as garages, cellars, stairs and laundry rooms, are located along the north side of a building so that they serve as protective buffer spaces. Basements and attics can perform similar functions. For a buffer space of this type to be effective, insulation should normally be placed between it and the heated part of the building rather than on the outside of the buffer space. Unheated sunspaces placed along the south side of the building also serve as buffer spaces - but of a different type. Their function is also to accumulate heat during the day using the greenhouse effect.

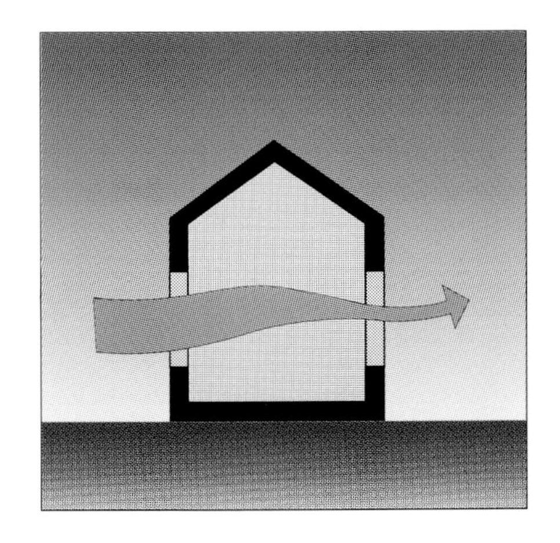

CHAPTER 5

THE BUILDING COOLING STRATEGY

In warm periods of the year, the sun's rays, infiltration of hot outside air into the building and internal gains from the activities of occupants and from equipment can all lead to an unacceptable. To achieve a comfortable internal temperature, a number of measures should be taken:

• **Solar control:** to prevent the sun's rays from reaching and entering the building.
• **External Gains:** to prevent increases in heat due to conduction through the building skin or by the infiltration of external hot air.
• **Internal Gains:** to prevent unwanted heat from occupants and equipment raising the internal temperatures.
• **Ventilation:** unwanted hot air may be expelled and replaced by fresh external air at a suitable temperature.
• **Natural Cooling:** to transfer excess heat from the building to ambient heat sinks.

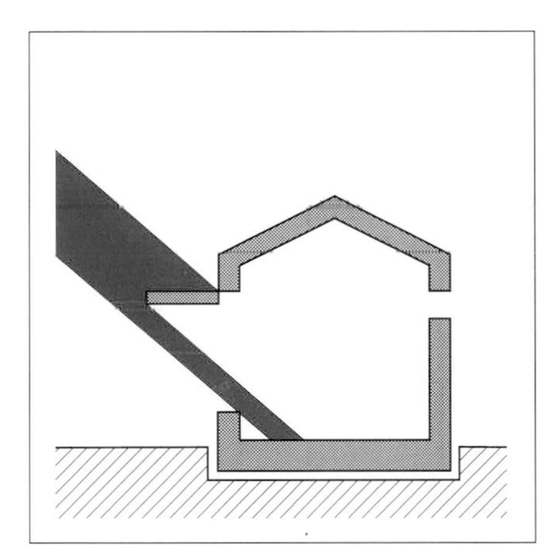

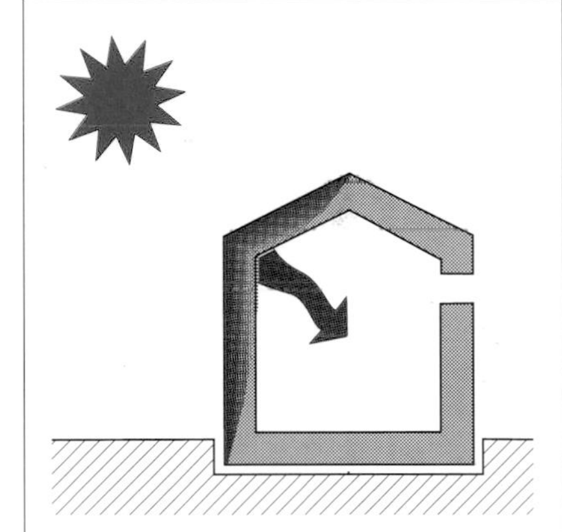

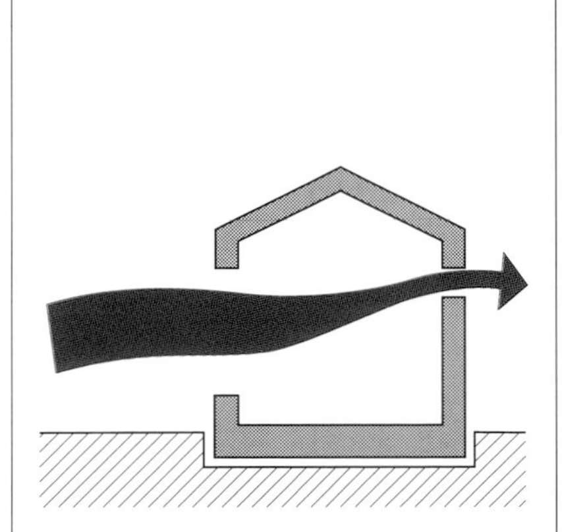

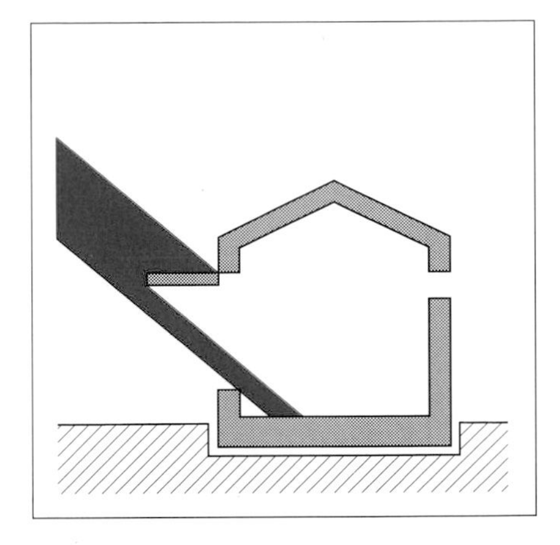

COOLING STRATEGY
SOLAR CONTROL

At certain times of the year in both temperate and warm climates, solar gains through glazing can be excessive, creating uncomfortable increases in temperature. This may be controlled by preventing the sun's rays from reaching the interior. Permanent, movable or seasonal features can be used. Among the commonly-used examples are overhangs, awnings, movable blinds and planted screens.

SHADING

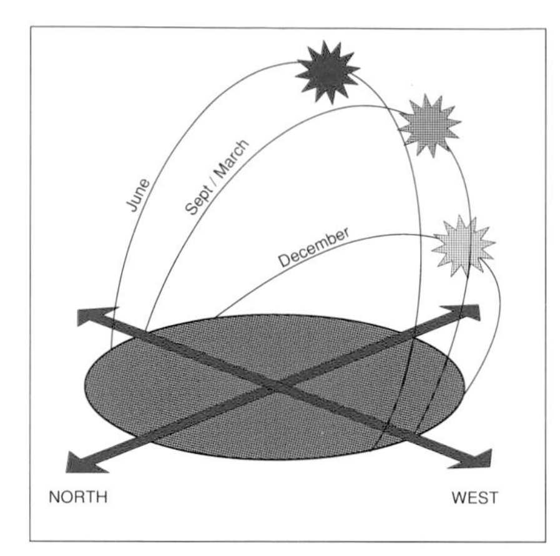

The most efficient way of protecting a building is to shade its windows and other apertures from unwanted direct sunlight. The degree and type of shade necessary depends on the position of the sun and the position and geometry of the part of the building being shaded.

In summer, when the solar beam falls on the south side of a building the sun is fairly high in the sky. Thus any south facing apertures receive less solar radiation and it is easy to protect them. Shading of east and west facing windows, on the other hand, poses a greater problem because the sun is low in the sky when it is in the east or west and a greater amount of solar radiation reaches these windows. One solution, therefore, is to consider reducing as far as possible the area of east and west-facing glazing.

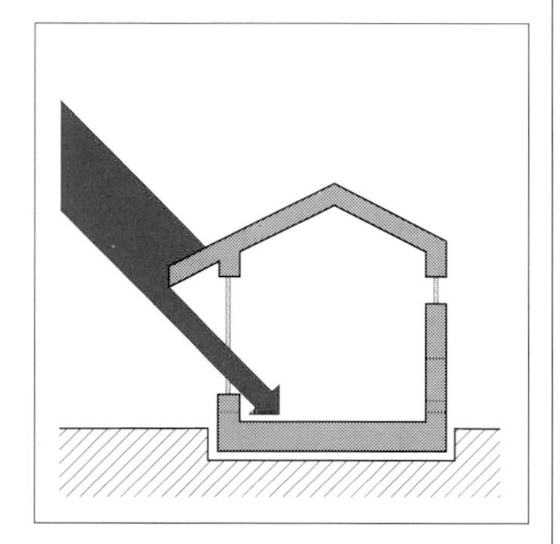

FIXED SHADING

The design of a shading device which is a fixed part of the building, as distinct from one which can be adjusted, must take into account the orientation of the aperture being protected.

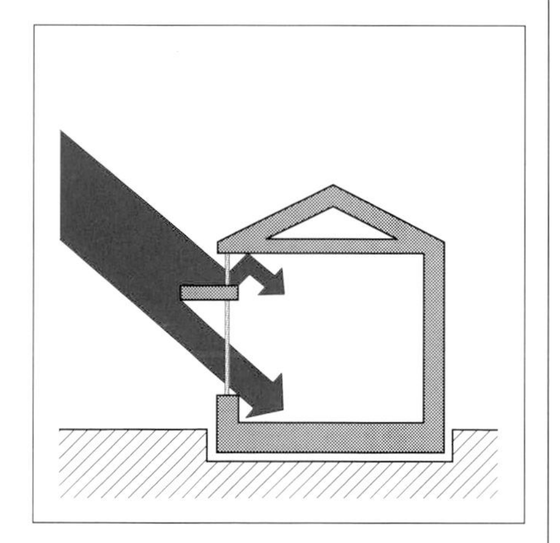

South facing windows can be shaded by an overhang above the glazed element. To obtain the maximum benefit from the sun's rays in winter, when they can make a useful contribution to heating requirements, it is sensible to locate the overhang in such a position that rays can pass through the opening when the sun is low in the sky. The depth of the overhang should take into account not only its distance above the window but also the aperture height. The length of the overhang is determined by the window width.

East and west facing windows can benefit from lateral shading. Because the sun's position changes, a movable vertical screen can be the most effective way of providing this although it can pose problems of stability and maintenance. If a fixed screen is to be used, its dimensions should be determined from the width and height of the window and the distance of the screen from it. It can be useful to use sun path diagrams to estimate the amount of shadow thrown onto the window by the proposed screen.

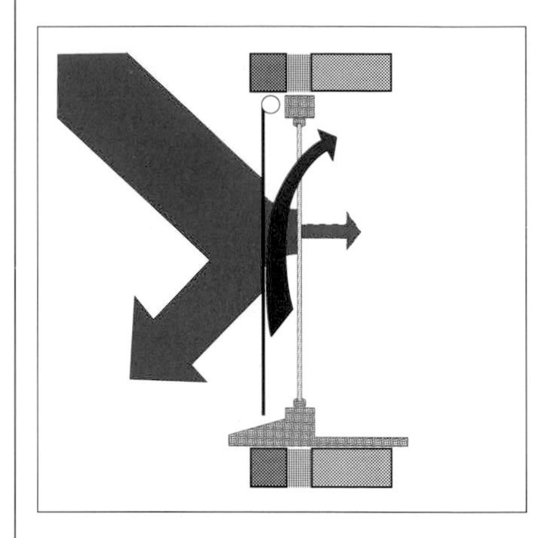

ADJUSTABLE SHADING

The effectiveness of fixed screens varies according to the seasonal changes in the position of the sun. Because climatic seasons do not correspond to solar seasons it is preferable in regions with a long heating season to erect movable protection which can be adjusted easily. Shutters, blinds, venetian blinds, awnings and curtains are all examples of adjustable shading devices. They can also be used in winter to increase thermal insulation. The effectiveness of their shading is expressed by a shading coefficient, the ratio of the solar energy passing through a protected opening to the energy which would pass through the opening if it was unprotected. Usually, a simple window is taken as the reference.

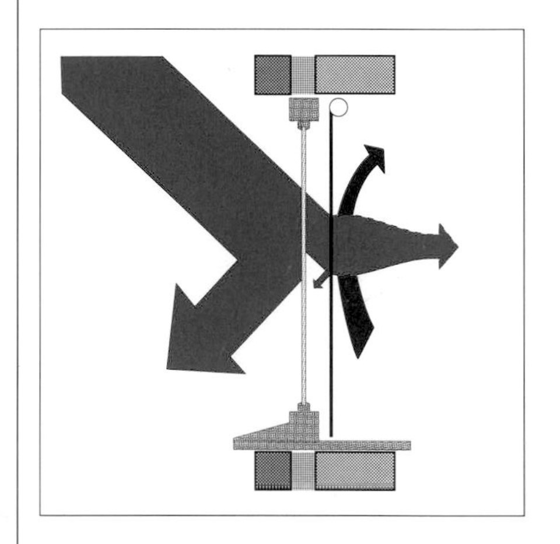

In designing a movable shading system, the aim should be to minimize unwanted solar gains but not to darken living space and force occupants to use artificial lighting. It is preferable to place screens outside rather than inside the building so that most of the sunlight can be reflected before it reaches the glazing. Unless they are reflectant, internal screens do not stop the sun's rays until after they have passed through the window and so both the air between the window and the screen and the screen itself heat up. The screen can be sandwiched between the two layers of glass in a double glazed unit, thus enabling the features of an external screen to be retained while avoiding maintenance problems. The colour and surface condition of a screen also play a role in determining its effectiveness. Their reflection and absorption properties have an effect on the amount of solar radiation entering the building.

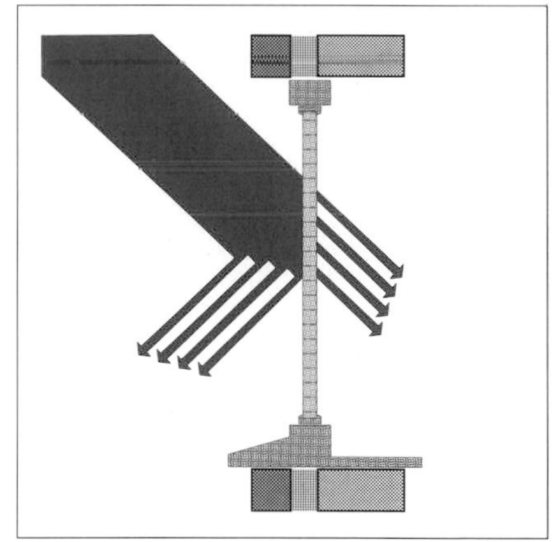

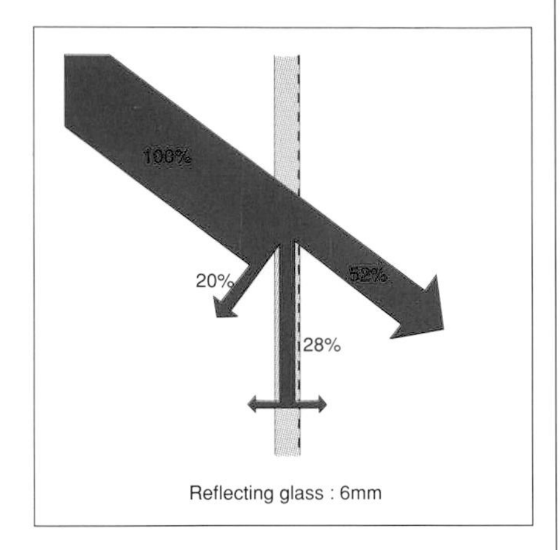

Reflecting glass : 6mm

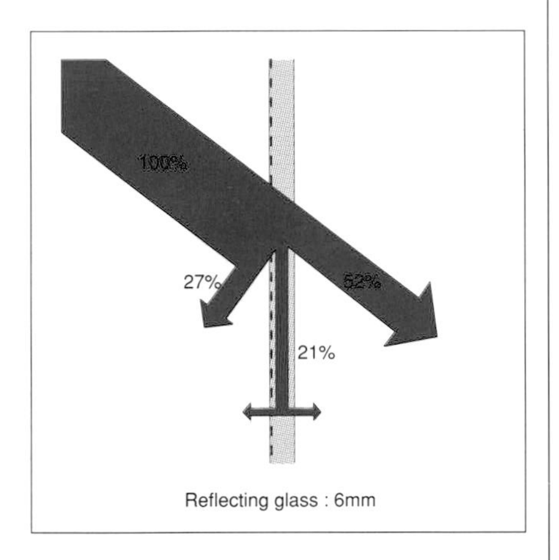

Reflecting glass : 6mm

SPECIAL GLAZING

For windows which are difficult to shade, special glazing can be useful in preventing unwanted heat gains while retaining the view and light.

Absorbing glass reduces the overall transmission of solar radiation through the window by cutting down on direct transmission and increasing re-emission towards the exterior after absorption.

Reflecting glass is made by coating the glass with a thin layer of highly reflective metal oxide. Ideally, this should be on the outside of the window. This would, however, create durability problems and the film is as a result normally placed on the inside face of the outer layer of glass or the outside face of the inner layer. If the reflective layer is placed on the inner glazing, care must be taken with the design of the double glazed unit to ensure that the air in the gap cannot heat up and thus cause loss of seal.

Absorbing or reflecting glasses are recommended mainly for windows facing east or west. They are rarely used in domestic buildings at present but have good potential for the future, especially in hot climates.

In addition, while not yet commercial, there is considerable development in photochromic, thermochromic and electrochromic glasses which modify the incoming rays of the sun so that the optical properties of the glass change and, for instance, darkening of the glass can occur.

The diagrams shown here give typical values for the proportions of solar radiation transmitted by a range of glazing systems.

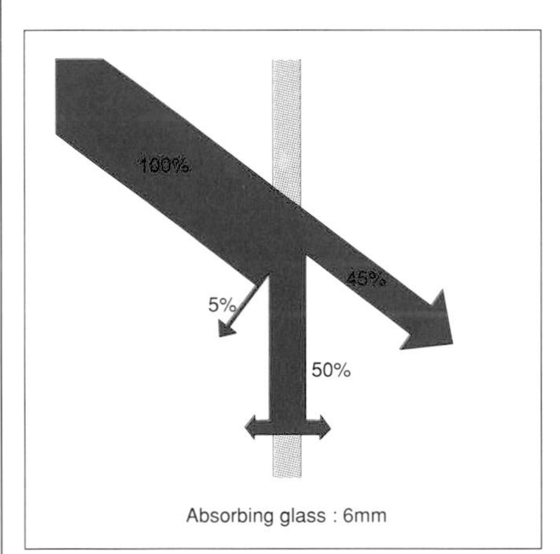

Absorbing glass : 6mm

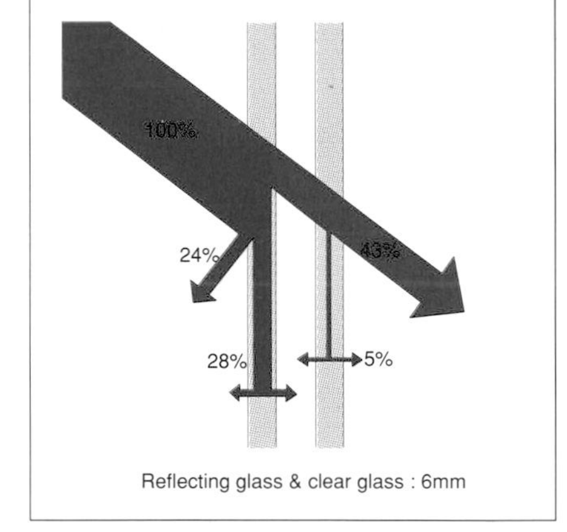

Reflecting glass & clear glass : 6mm

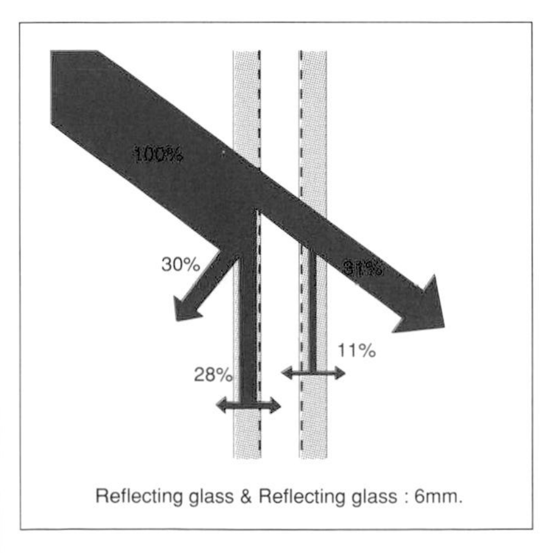

Reflecting glass & Reflecting glass : 6mm.

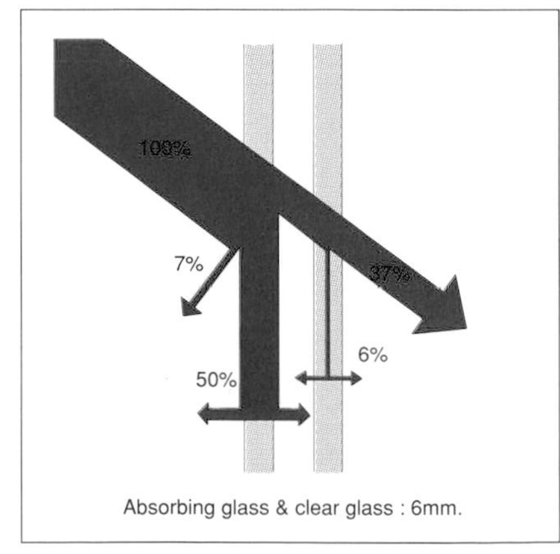

Absorbing glass & clear glass : 6mm.

SHADING BY NEIGHBOURING BUILDINGS, TOPOGRAPHY AND PLANTING

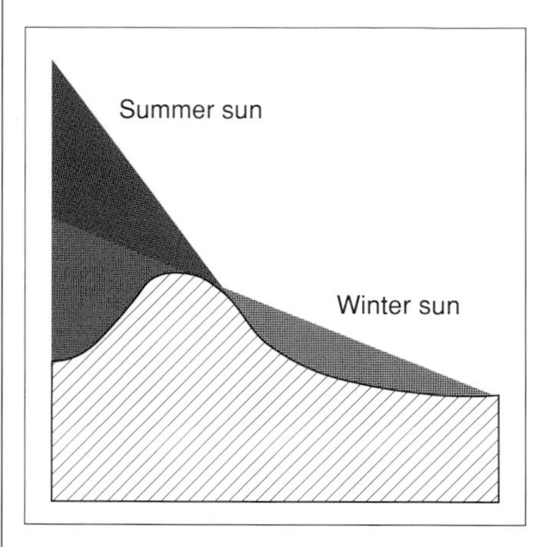

Almost invariably, some shade will be thrown on a building by neighbouring construction. This effect is frequently put to deliberate use in hot, dry climates where towns may be planned and built in a very compact form with narrow streets so that all the buildings are shaded to some extent from the sun. In such situations, however, it is important that the buildings are not put so close together that ventilation is difficult to achieve. Even in the design of individual buildings and their surrounding spaces, it is possible to use existing neighbouring buildings to block off unwanted solar radiation.

The topography of a place, too, can often create shade. Therefore, when choosing a location for the building on a site in regions where overheating is likely, it is sensible to try to take advantage of this and construct the building in the most shaded area. The shadow cast by the topography will be a function of the sun path, the orientation and the tilt of the land.

Shading can also be provided by vegetation. If the planting is deciduous, a certain amount of advantage can be taken of the sun's rays in winter when the branches are bare but a progressively increasing amount of shade is created from spring onwards as the leaves grow. It is best to choose plants with dense foliage but few branches so that maximum protection is provided in summer and minimum shade in winter. It should be noted that a bare tree will block out some 20-40% of the sun's rays.

COOLING STRATEGY
EXTERNAL GAINS

In hot climates, walls and roofs are heated by the sun and by the warm outside air. This may produce uncomfortable conditions inside. Gains can be minimized by insulation, reduced window size, use of thermal inertia, reflection, or compact site layout. Infiltration gains can be reduced by cooling the incoming air and by reducing infiltration.

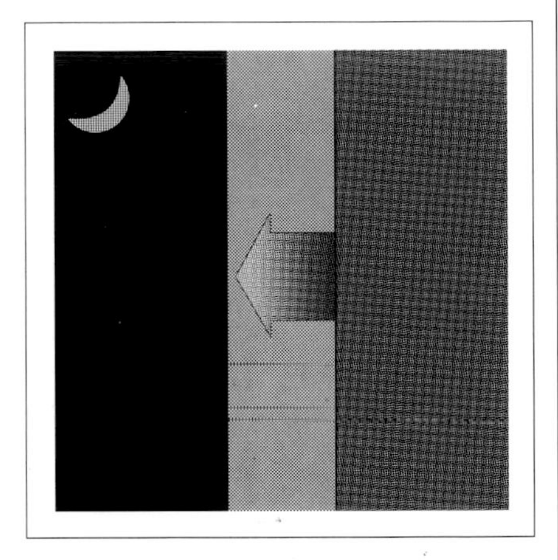

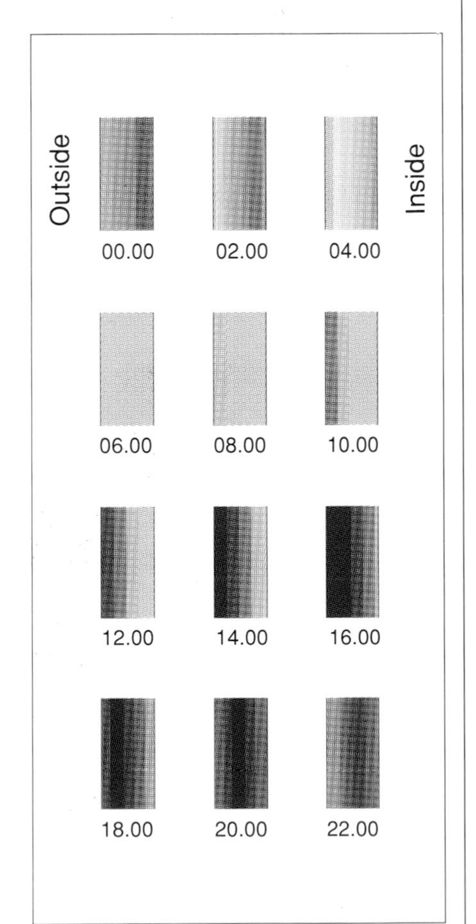

EXTERNAL GAINS

TRANSMISSION GAINS

Although the concept of reducing heat flow through a building envelope by increasing the insulation level is most commonly applied to heat conservation in the heating season in cooler regions of Europe, it can also be used to prevent overheating by conduction in summer in hot climates. In summer, of course, the heat flow is from the outside to the inside of the building - the opposite direction of the winter heat flow. A discussion of the principles of insulation can be found in the section on heat conservation earlier in this text.

In addition to insulation, there are three other ways of reducing heat flow through the envelope. The first makes use of the thermal inertia of the building envelope. The second involves provision of a barrier to reflect radiation away from the building. In the third, the surface area of the building envelope is reduced by means of a compact layout.

THERMAL INERTIA

This method is based on the fact that there is a time delay due to the thermal inertia of the walls and roof, etc., in the flow of heat through the building envelope which can be exploited in a heavyweight building for cooling purposes. The concept is particularly helpful where there are significant daily variations in external temperature - in hot, dry climates, for instance.

When solar radiation strikes an opaque surface such as a wall or a roof the exterior surface absorbs part of the radiation and converts it to heat. Part of the heat is directly re-emitted to the outside. The remainder is conducted through the wall or roof at a rate which depends on the thermal diffusion characteristics of the material.

When the temperature of the exterior surface drops because of a fall in ambient temperature part of the stored heat is emitted outside.

At night, the air temperature inside the building is higher than the temperature outside. The heat flow to the outside therefore continues and the temperature of the wall or roof continues to decrease, thus eventually cooling the interior.

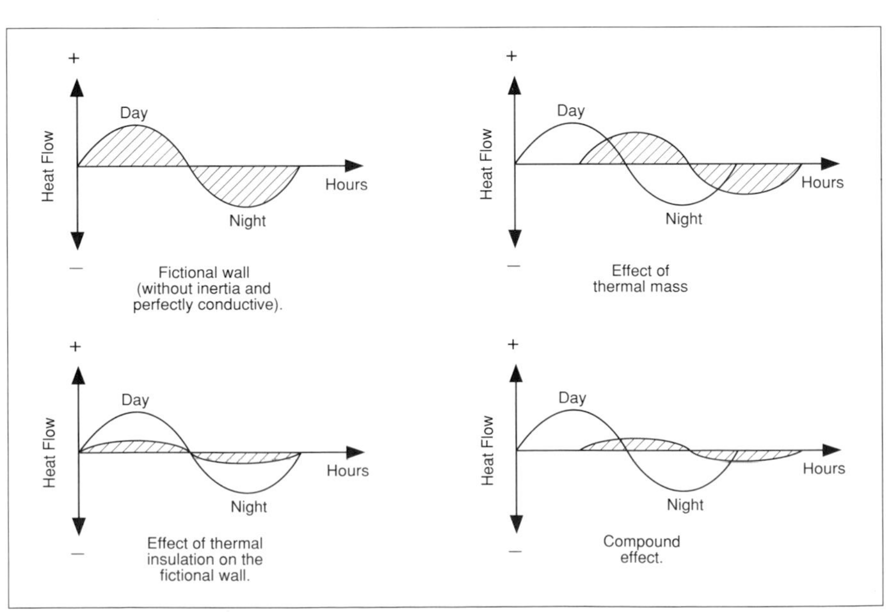

REFLECTION

Light colours have the property of reflecting short wave, solar radiation and it is for this reason that buildings in hot climates are often painted white.

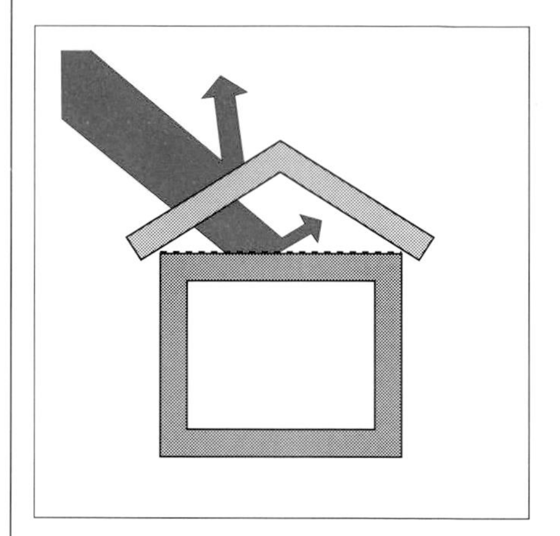

In an air-filled cavity wall or roof space where the air is still and convection is therefore low, thermal radiation is the prime mechanism for transfer of heat. The thermal radiation can be reflected away from the occupied part of the building by coating the face of the partition separating the cavity from the occupied areas with a highly reflective material such as aluminium foil. Such a foil also increases the thermal resistance of an insulation layer if it is placed adjacent to the insulation with a small air space between the two.

Radiation barriers are recommended for those parts of lightweight buildings in hot and humid climates where it is difficult to provide protection from the heat. They are particularly effective in places where the heat flow is downwards, as in an attic in summer. Reduction of heat transmission can be as much as 90% when a simple reflective sheet is placed on the floor of an attic.

It is however important to remember that where insulation is also used there may be a risk of condensation in winter when the heat flow is reversed.

SITE PLANNING

The very compact urban layout sometimes found in hot, dry climates can help to maintain cooler temperatures in and around buildings. Compactly designed building have smaller surface areas and this helps to reduce transmission losses. Dense urban planning also allows buildings to benefit from mutual shading.

INFILTRATION GAINS

When diurnal temperature variations are significant (as they are, for instance, in hot, dry climates) unwanted air infiltration should, if possible, be avoided during the day and the building ventilated at night so that any stored heat can be released from the thermal masses. In addition, steps should be taken to create a cool area round the building so as to reduce the temperature of the infiltration air and, where appropriate, to increase ground contact.

COOLING STRATEGY
INTERNAL GAINS

Artificial lighting, appliances and the activities of occupants all lead to internal casual heat gains. In hot climates these gains must be minimized. This is achieved by daylighting where appropriate and by use of high-efficiency artificial lighting, by accurate control, by choosing efficient appliances which generate little heat, and by expelling the heat generated.

	kWh/day
Occupants	4.0
Lighting	1.5
Appliances and Cooling	6.5
Hot water	3.0
Total	15.0

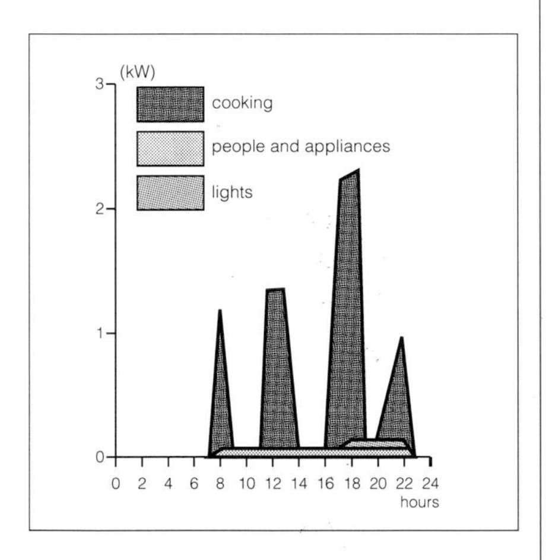

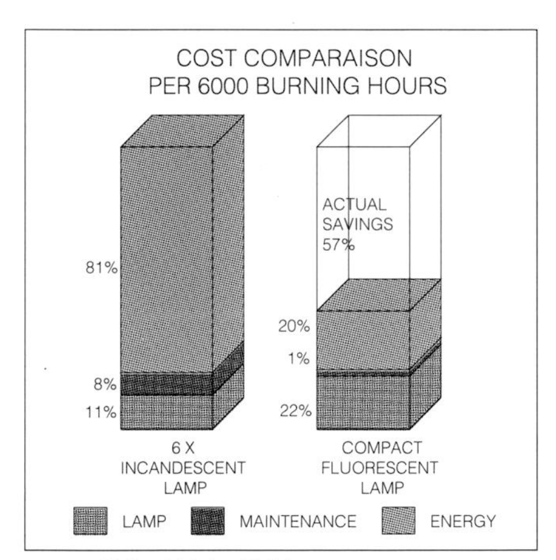

source: Philips Lighting Division

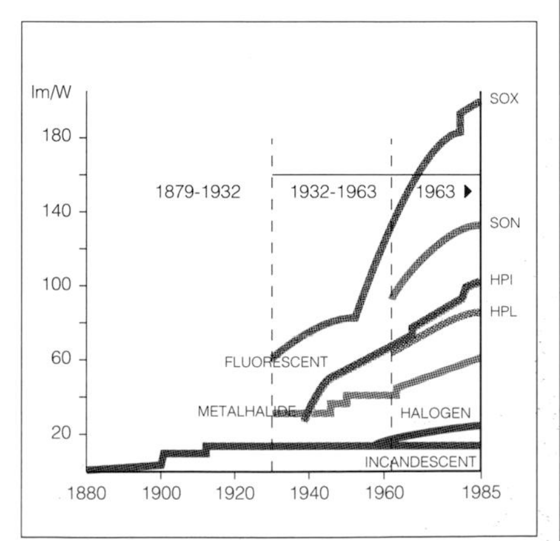

Source: Philips Lighting Division

SOX = Low pressure sodium
SON = High pressure sodium
HPI = Metal halide high pressure gas discharge
HPL = High pressure mercury

INTERNAL GAINS

The distribution of internal gains throughout a building will vary considerably from one building to another according to its occupancy and type of use. Even in the domestic sector, the pattern generated by one family will differ from that of another. At the design stage, therefore, it is difficult for the architect to know precisely what the internal gains of a building will be. It is, however, possible to make an assessment based on standard profiles for room and dwelling types for a typical family.

The table opposite shows typical incidental heat gains per day for a house with three bedrooms.

The second figure opposite gives an indication of the internal heat gain in a well-insulated north-facing kitchen on a dull January day.

ARTIFICIAL LIGHTING

When electricity is used for lighting, not all of the energy is actually transformed into light. Some of it (the proportion depends on the efficiency of the lighting equipment) is converted into heat and dissipated into the room. It is sensible, therefore, to reduce these gains by applying the strategies for natural lighting found in the daylighting section of this book This is particularly important in those buildings where air conditioning might otherwise be necessary It is always important to use high-efficiency lighting devices to maximize the proportion of energy transformed into light. In addition, artificial lighting can be zoned and controlled so that lights are only activated in areas where and when there is a need for light and the daylighting is inadequate. The intensity of the artificial lighting can also be altered automatically in response to the available daylight.

The choice and placement of lights fittings will depend on the type of space to be lit and the activities carried on there. In the home, for example, a change to compact fluorescent replacements for conventional incandescent bulbs can provide worthwile energy savings, as shown in the diagram opposite. The life of compact fluorescent bulbs is typically eight times longer than incandescent bulbs.

In commercial and industrial buildings a wide range of energy efficient lighting options exists, from high frequency fluorescent lighting to high and low pressure gas discharge lamps. Very significant energy savings can be made by the correct choice of artificial lighting systems and automatic lighting control can extend these savings.

APPLIANCES AND ELECTRICAL EQUIPMENT

Other electrical equipment, apart from lighting, gives off heat when in use. To minimize internal gains of this type, domestic appliances such as refrigerators and washing machines and other equipment (computers, photocopying equipment and so on) should be selected which are efficient in energy use so that they release as little heat as possible. Good, energy-efficient appliances can cut electricity bills by half. Where it is appropriate and possible, such equipment should be located together on a side of the building which is opposite to the prevailing wind so that the infiltration air does not cause the heat given off to flow through the whole building.

OCCUPANTS

Everyone, depending on his metabolism and level of activity, gives off what is at times a significant quantity of heat. When several people are in a room together, therefore, the heat gains due to occupants can on occasion be excessive. To minimize these gains, it is important for the architect to bear in mind at the design stage the possible occupancy level of each room and the range of activities to be carried out there, ensuring, for instance, adequate ventilation in densely-occupied rooms. In addition, it can be helpful in hot climates to arrange external spaces adjacent to high-occupancy rooms so that there can be an overspill into them when the internal gains in the building become too high. Attention should be paid to shading and planting these external spaces so that they are cool and attractive.

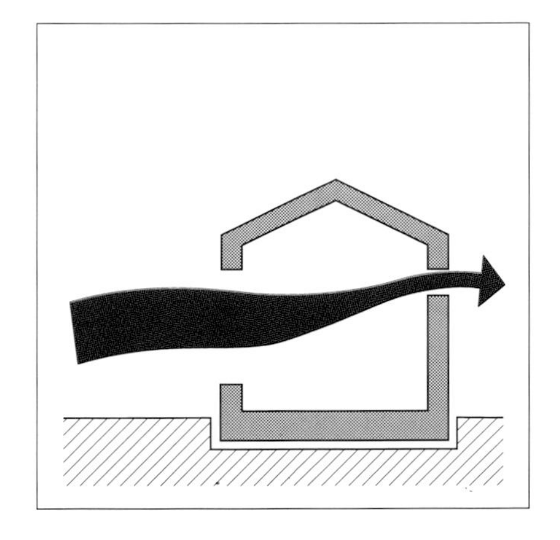

COOLING STRATEGY VENTILATION

Even when steps have been taken to shade a building, to reduce heat gains and to minimize the flow of external warm air into the building, internal temperatures in hot climates during summer can often be higher than those outside. Ventilation using cooler fresh air driven through the building by naturally occurring differences in wind or air pressure can help to remedy this problem.

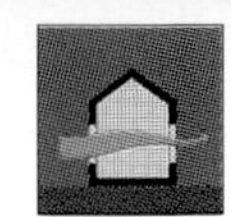

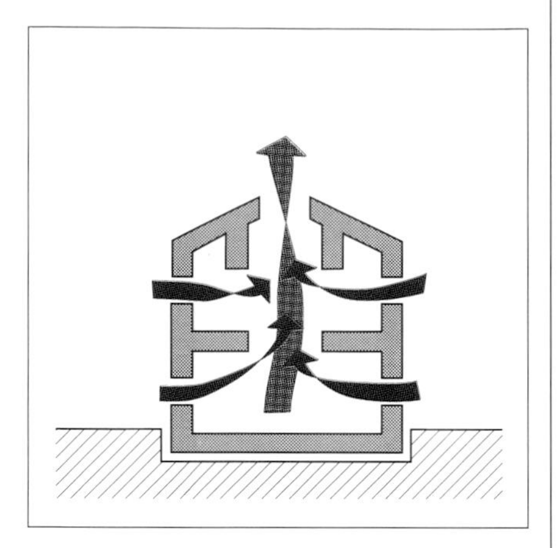

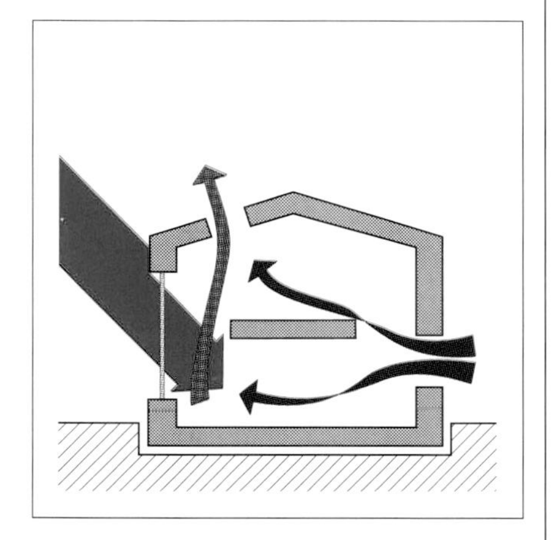

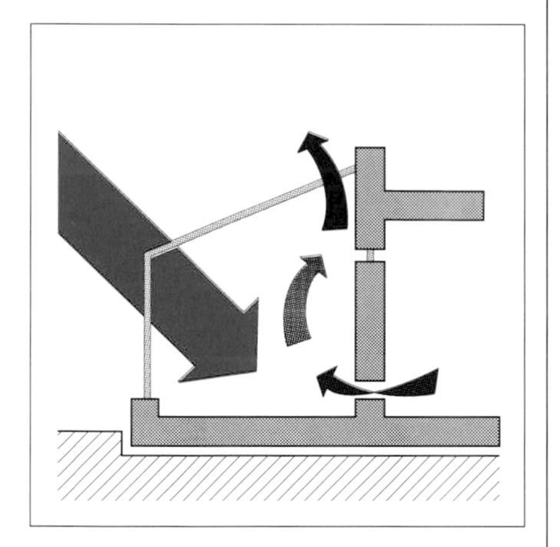

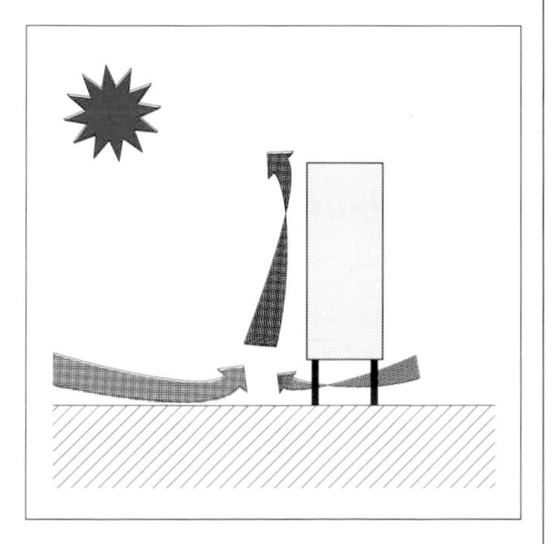

TEMPERATURE GRADIENT EFFECT

The general principles of circulation using temperature gradients have already been described earlier in this text (see section on thermocirculation in the Heating Strategy chapter). Briefly, when two air masses have different temperatures, their densities and pressures are also different and this gives rise to movement of air from the denser (cooler) zone to the less dense (warmer) one. In situations where the air inside a building is warmer than ambient air and cooling is required, the temperature gradient effect can be used to expel the warm air from the building.

One way of doing this is to invoke the stack or chimney effect by providing openings at the top and bottom of the building. The warm air will rise naturally and escape from the top outlet while cooled fresh air will enter through the openings at the base. The most thorough ventilation is achieved when the openings are placed vertically.

The chimney or stack effect can also be put to effective use to release unwanted heat from a building via a sunspace or atrium. The air movement in the sunspace can give rise to an intake of air from the building into the sunspace. This may in turn be dissipated to the outside. The same effect can be used to create cross ventilation. Here, air enters the building from cool pockets created in outside spaces on the building's shady side and flows through the building towards the sunny side, where it rises as it is exposed to the sun.

WIND PRESSURE EFFECT

It is possible to increase the dissipation of heat from a building using the wind pressure effect. When wind strikes a building a high pressure on the exposed side and a low pressure on the opposite, sheltered face results. Usually, the speed and direction of local winds are variable. At an individual site however, a building can often be positioned in relation to neighbouring buildings, planted vegetation and other obstacles so that a wind is induced in a known constant direction at a reasonably steady rate. The conditions for ventilation are best when the wind strikes the building at an angle of up to 45 degrees.

The movement of air across a site is from high pressure zones to low pressure zones, through openings in the building envelope. The size and position of the openings determines the speed and direction of air movement within the building. The air speed is greatest when openings by which air leaves the building are bigger than air inlet openings, however inlet openings must be of adequate area. The best distribution of fresh air throughout the building is achieved when the openings are diagonally opposite each other and air flow is not hindered excessively by partitions and furniture, etc.

Maximum ventilation should be provided during the day in occupied areas of the building at head height. In addition, there should always be a good flow of fresh air along the building's most massive elements so that as much heat as possible is dissipated from them.

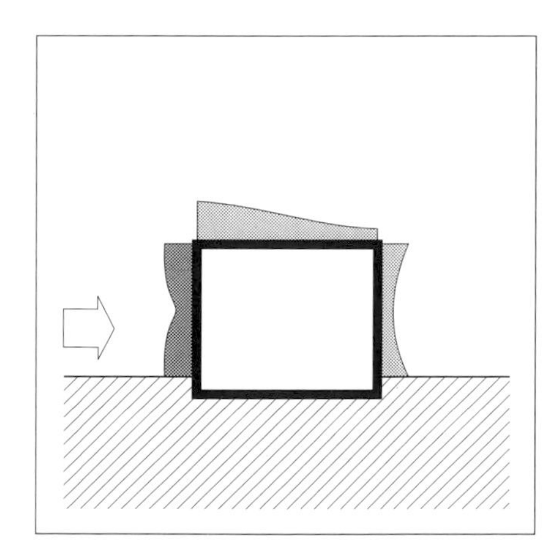

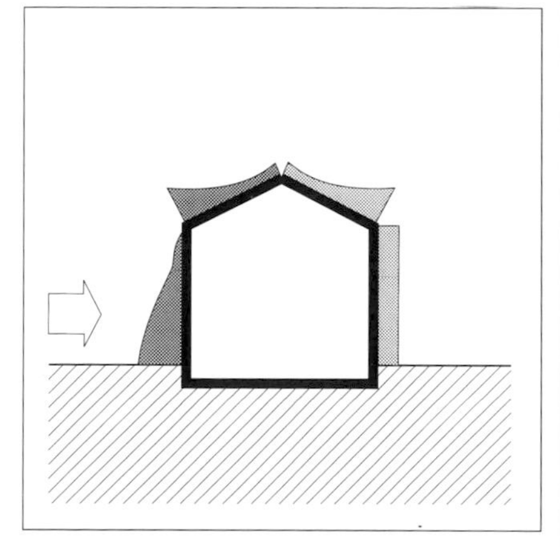

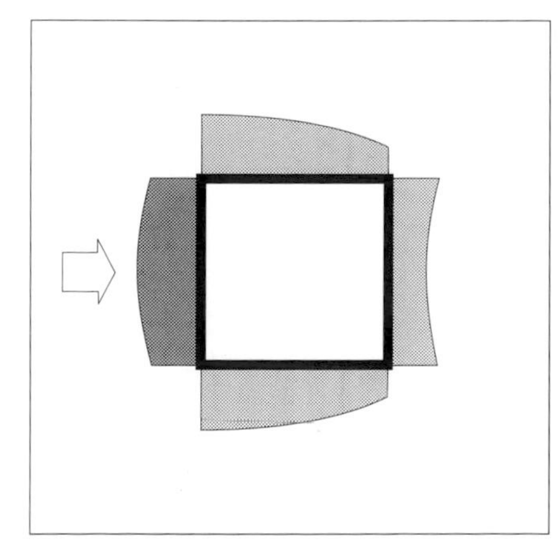

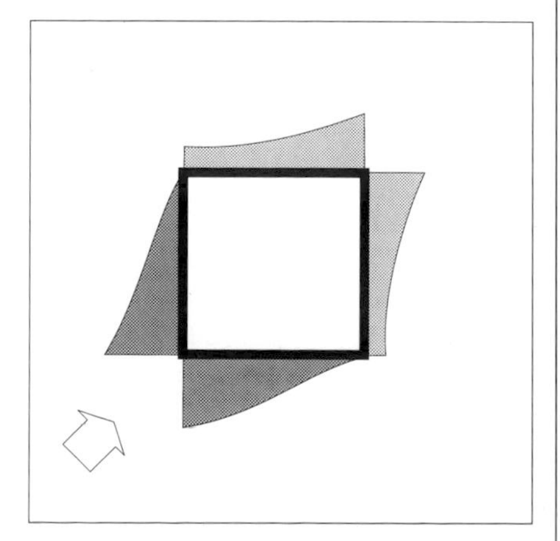

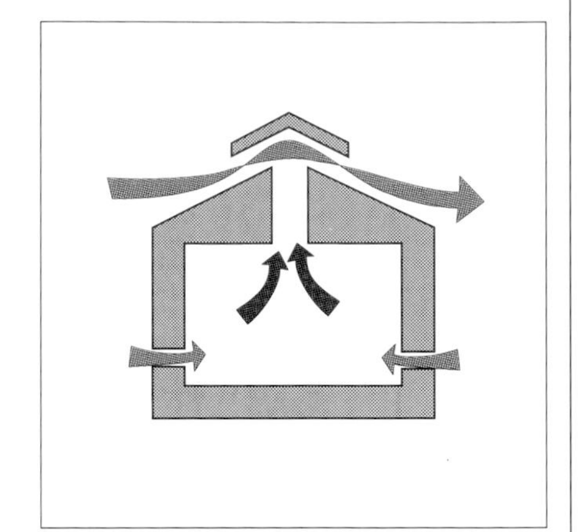

External wind deflectors can be incorporated in the design of the building to alter the pressure at openings so that certain zones can be preferentially ventilated by specially induced air flows. Correct positioning of the deflectors requires prior analysis of local winds.

The Venturi effect can also be used to induce circulation of air in a particular direction. The air is encouraged to flow through a constricted part of the building. At this position, its speed increases and the pressure decreases accordingly. The reduced pressure creates an air flow which can be used to drive hot air from the building and thus cause ventilation.

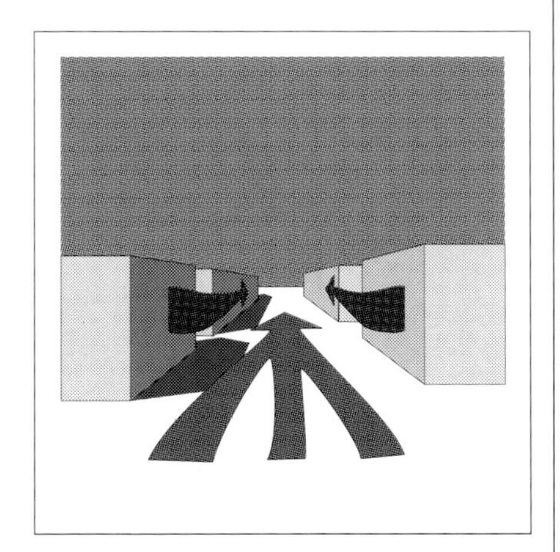

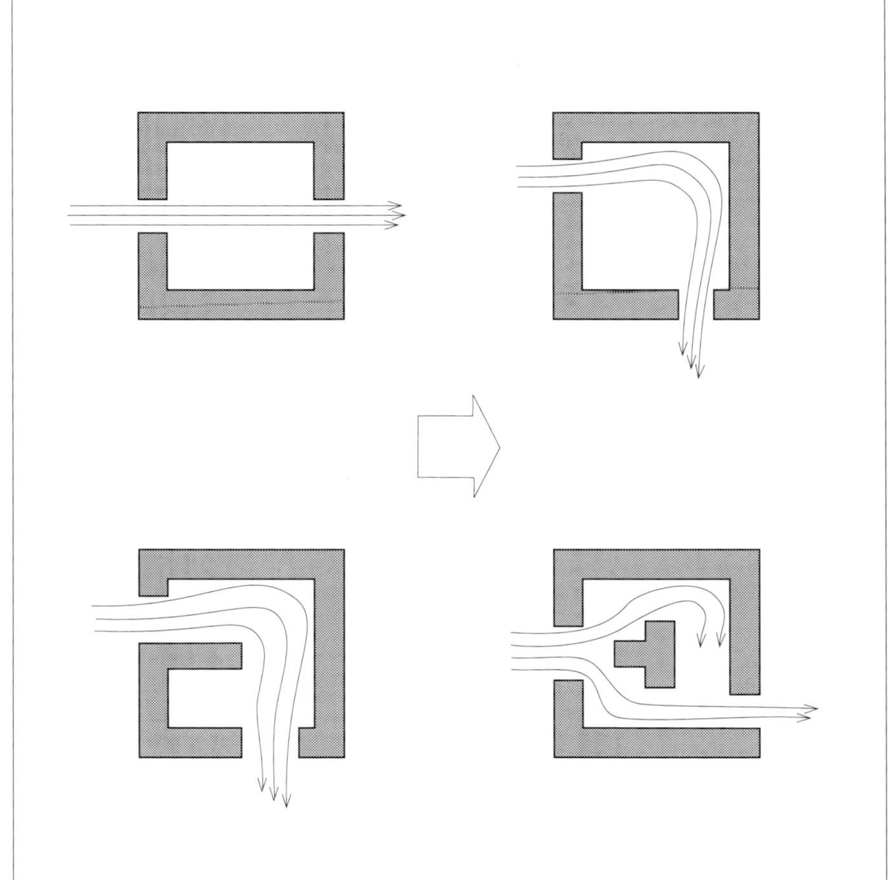

When it is not possible to put openings in positions appropriate for good ventilation, the wind can be funnelled around the building to a convenient position using fencing, walls, hedges and shrubbery.

In towns, a number of wind effects can be encountered resulting from the relative position of existing buildings to the wind flow. As a result it can be possible to locate a new building so that the wind channelling provides good ventilation flows through the building and causes cooling. Care must be taken when using this effect because, of course, it can produce undesirable cooling and other unpleasant results in winter.

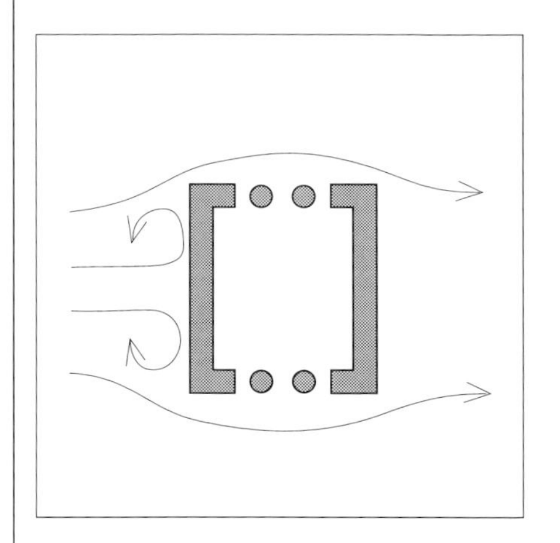

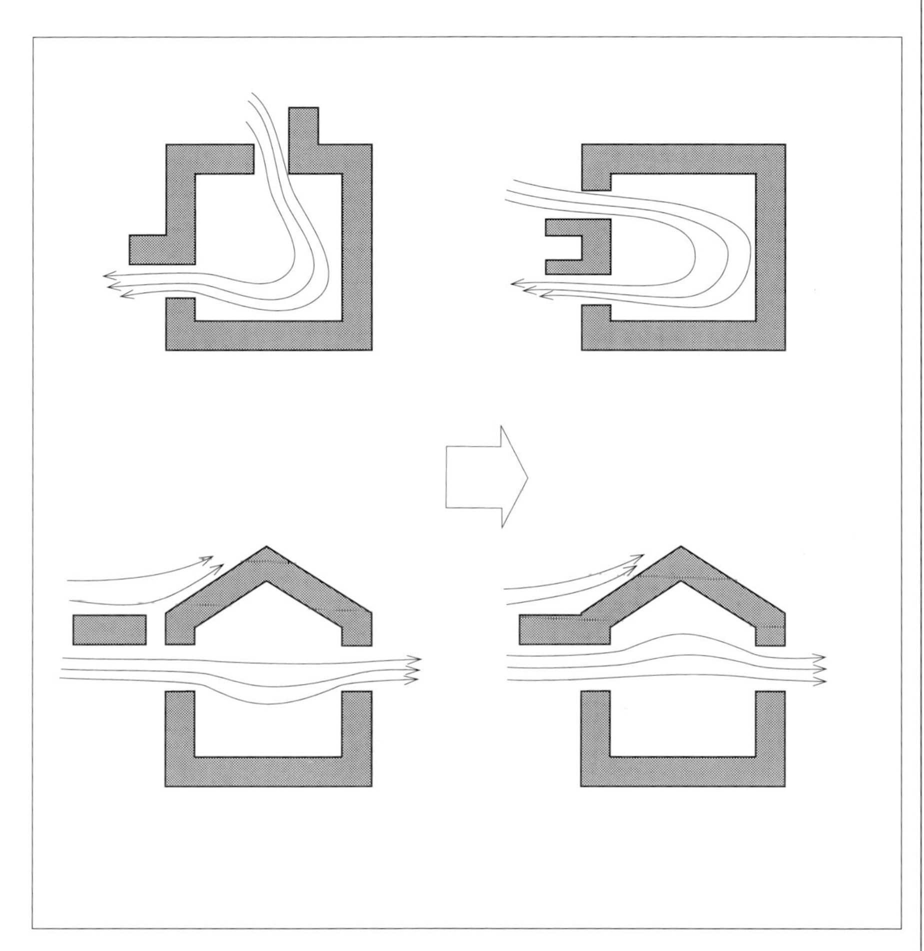

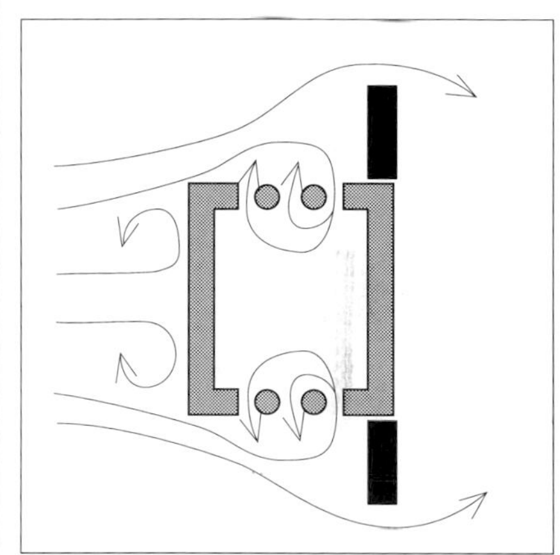

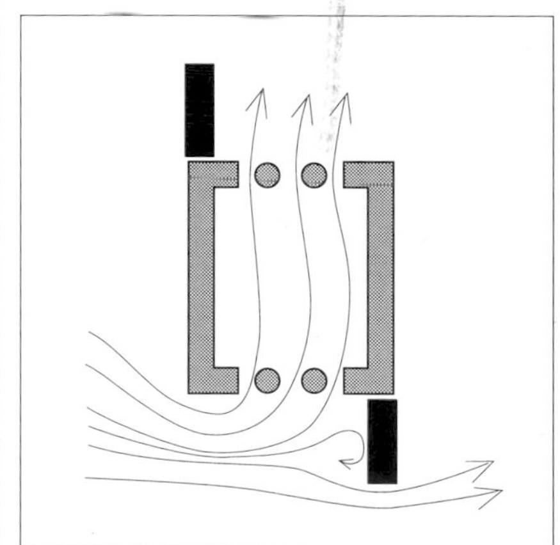

COOLING STRATEGY
NATURAL COOLING

The final step in a cooling strategy is to provide cooling by natural means. Internal air speeds can be increased to maximise perceived cooling.Air adjacent to the building can be cooled by evaporation. The temperature of ventilation air can be reduced by ground cooling. The building can be cooled by night-time radiative heat loss to the sky and by cross ventilation.

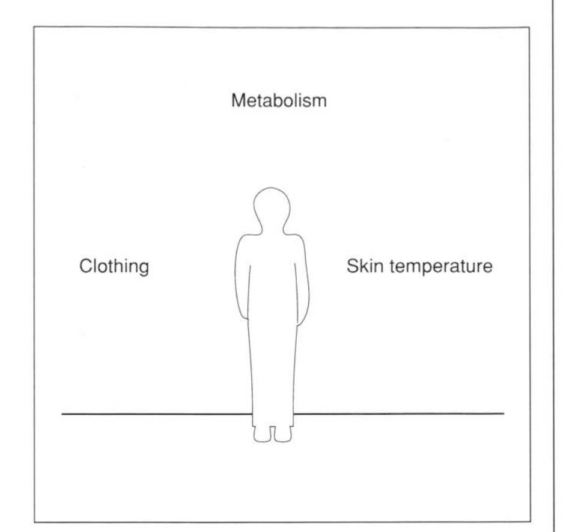

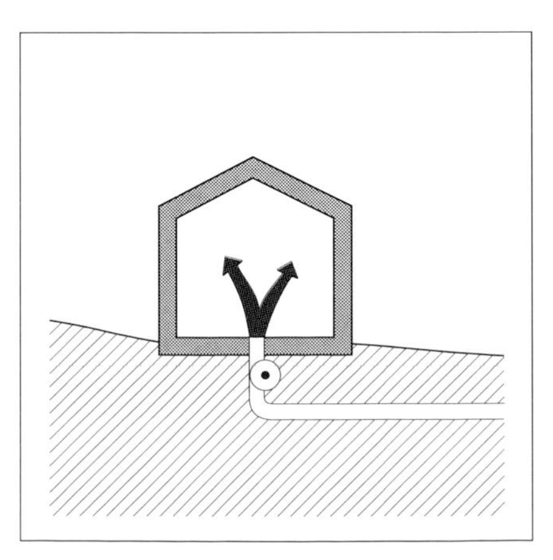

OCCUPANTS' PERCEIVED LEVEL OF COOLING

Building occupants' perceptions of comfort are influenced by a number of parameters. Some of these such as air temperature, mean radiant temperature, relative humidity and air velocity relate to the environment. Others relate to the occupants and include, for instance, activity level, clothing and skin temperature.

Identical levels of comfort can be achieved by various combinations of these parameters. An alteration in one or several of these can, therefore, lead to an improvement in the occupants' perceived level of cooling. Under certain circumstances, it is possible to change one of the variables in such a way that comfort is increased without affecting the energy balance in the room.

A reasonable rise in the speed of air in the room, for instance, can produce increased comfort provided the temperature of the air is lower than skin temperature. This is because the flow of air causes heat loss by convection.

Perceived cooling can also be produced by increasing the rate of evaporation on the surface of the skin by creating an air movement to break up the layer of saturated air which surrounds the body.
These effects can be produced by natural ventilation. Failing that, of course, mechanical fans can be used.

COOLING OF INFILTRATION AIR

EVAPORATIVE COOLING

To change its state from liquid to vapour, water requires a certain amount of heat known as the latent heat of vaporization. When this heat is supplied by hot air there is a drop in air temperature - accompanied, of course, by an increase in humidity.
The evaporative cooling effect can be maximized by increasing both the air/water contact area and the relative movement of the air and water.
The effect can be put to good use by establishing pools, fountains and water jets, etc., in outdoor spaces next to buildings to cool the ventilation air before it enters the building. Evaporative cooling cannot be used in humid climates where the air is already close to saturation.

GROUND COOLING

Because the temperature of the ground below a certain depth is colder than that of the ambient air and remains at a fairly constant temperature year round, air for ventilating a building can be cooled by passage through an underground duct. The cooling process is one of convection coupled with evaporation if the ground is damp. The temperature drop depends not only on the ground temperature but also on the surface area of the duct. It may therefore be necessary to excavate a channel of some considerable length to obtain the amount of cooling required. If the channel is unlined or contains a pipe made of porous material, then problems could arise from decomposition of organic material in the channel or emission of pollutants from the soil.

In hot dry climates, the temperature of the ground below the surface is usually cooler than that of the air so that ground contact increases heat dissipation from the building. To make use of this idea, the parts of the envelope below the ground should not be insulated but should be waterproofed to avoid problems from moisture on their surfaces. However, this may be unsuitable where below-ground insulation is required for winter.

COOLING OF THE BUILDING ENVELOPE

RADIATIVE HEAT LOSS TO THE SKY

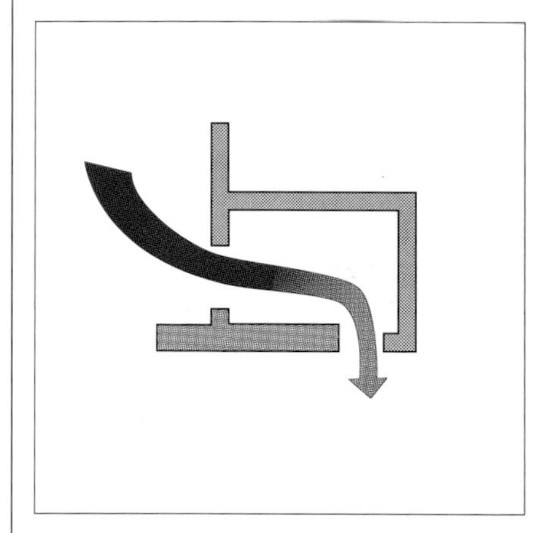

Radiative transfer of heat always occurs between two adjacent masses at different temperatures. Therefore, because clear night skies are (even in the warm season) invariably cold, a significant amount of the heat which has accumulated in a body of water or a building during the day will be radiated to the sky at night in clear weather. By the end of the night, the water or building will have cooled down significantly. The effect is less marked in humid climates because humid air is less transparent to infrared (IR) radiation than dry air.
The radiative cooling phenomenon is put to good use in roof ponds. In such systems, the heat accumulated in a building during the day is trapped and stored in the roof pond, which is protected on the outside by movable insulation. At night, the insulation is removed to allow the stored heat to be radiated towards the sky.

CROSS VENTILATION

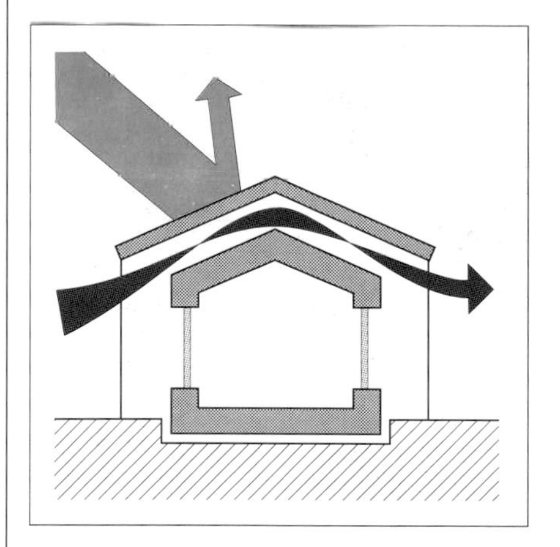

The rate of heat loss by convection from the building envelope can be accelerated by the wind. In a similar fashion, floors and ceilings in lightweight buildings can be kept cool by cross ventilation in the basement and attics. This procedure is particularly recommended for hot, humid climates.

In hot, dry climates where night-time temperatures are low, cross ventilation at night is an appropriate method of cooling. It is a necessary complement to heat storage. The night-time ventilation air is preferentially circulated past these high thermal inertia masses to remove the heat they have accumulated during the day.

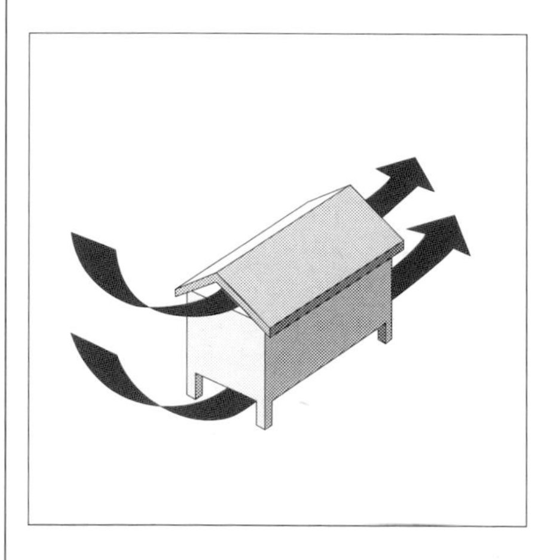

TEMPERATURE DIFFERENCES

INFILTRATION AIR

By taking pains over the building location and the arrangement of its surrounding spaces, it is possible to use natural cooling effects to reduce the temperature of the air round the building.
Topography and planted vegetation in the area, for instance, can trap cool air which has come down a valley at night and settled in depressions in the ground. The presence of water, too, can have a cooling effect.

Careful organization of rooms according to their function and thermal comfort requirements can help keep the living spaces as cool as possible. One way of achieving this is to face living areas north and use well-ventilated buffer spaces on the south side. An alternative approach is to create a cool core surrounded by well-ventilated spaces. This is exemplified by the courtyard house often found in warm climates.

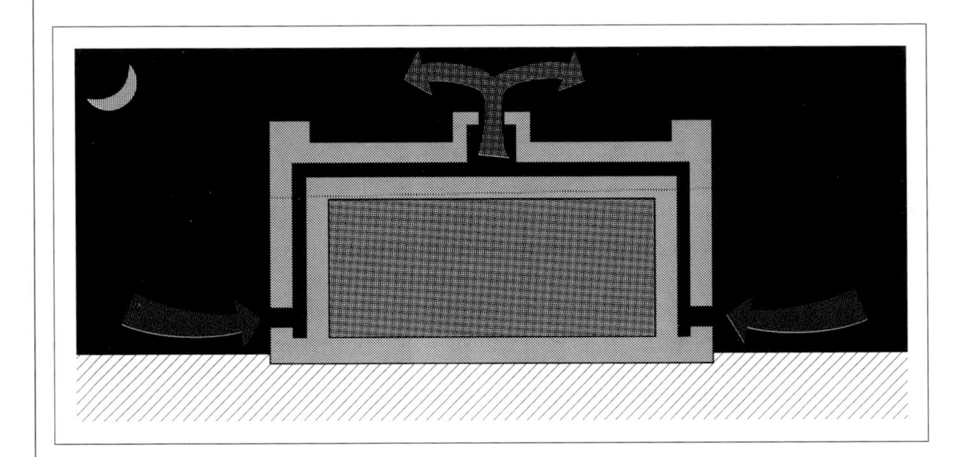

CHAPTER 6

THE BUILDING DAYLIGHTING

The optimal use of natural daylight, especially in buildings used mainly by day, can, by replacing artificial light, make a significant contribution to energy efficiency, visual comfort and the well-being of occupants. Such a strategy should include the potential for heat gain and conservation, energy savings by replacing artificial light and the more subjective benefits of natural light and external views enjoyed by the occupants.

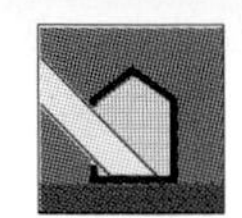

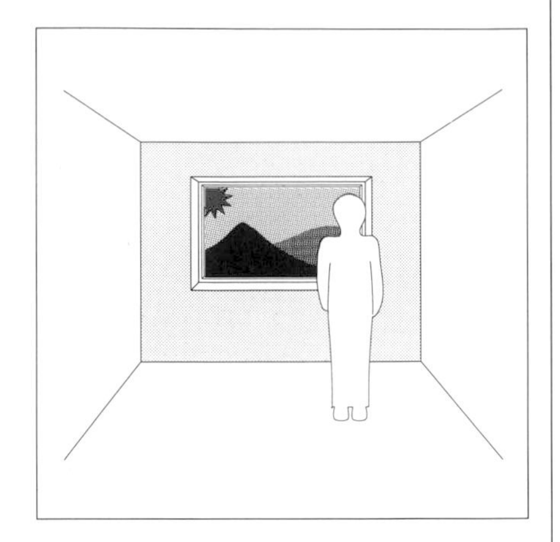

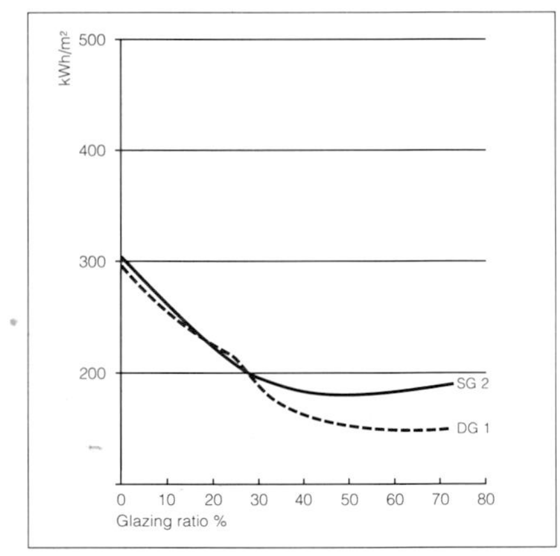

The graph above shows the overall yearly primary energy needs for one square meter of an office building oriented due south in London. It varies as a function of the glazed area percentage and of the type of glazing which is used (single or double glazing). An increase in the glazing leads to an important decrease in the energy needs: this is due to the increase in solar gains and to the reduction of the artificial lighting needs.

INTRODUCTION

Natural daylighting in buildings has much to recommend it. It has a variability and subtlety which is more pleasing than the relatively monotonous environment produced by artificial light. It helps to create optimum working conditions by bringing out the natural contrast and colour of objects. Windows and skylights give occupants contact with the outside world by offering an alternative long-distance view which is relaxing to the eyes after a lot of close work. The presence of natural light can bring a sense of well-being and awareness of the wider environment in which man lives. It is also claimed that exposure to natural light can have a beneficial effect on human health.

Visual comfort is the main determinant of lighting requirements and is also discussed later in this book in the section on Visual Comfort. Both the global illumination level and the light distribution in the room must be designed carefully, avoiding excessive contrasts in lighting levels and glare.
Daylighting is particularly important in commercial and other non-domestic buildings only used during the day. Even in these circumstances, however, it cannot, because of its uncertainty and variability, provide adequate illumination at all times. Artificial lighting systems must always be available to supplement natural lighting when this is necessary.

A good daylighting system has a number of elements, most of which must be incorporated into the building design at an early stage. This can be achieved by consideration of the following in relation to the incidence of daylight on the building:

- the orientation, space organization and geometry of the spaces to be lit;
- the location, form and dimensions of the openings through which daylight will pass;
- the location and surface properties of internal partitions which reflect the daylight and play a part in its distribution;
- the location, form and dimensions, etc., of movable or permanent devices which provide protection from too much light and glare;
- the light and thermal characteristics of the glazing materials.

ENERGY IMPLICATIONS

The proportion of the total energy required to run a building which is taken up by the artificial lighting system is much higher than might be thought. Case studies on an identical, conventionally designed and run, 54 square metre office room in Athens, London and Copenhagen indicate that in all three places artificial lighting typically represents an almost constant fraction of around 35% of the total lighting, heating and cooling costs throughout the year.

The pattern of artificial lighting use in a conventional building will, of course, differ according to latitude. In Athens, for instance, to protect the building from overheating due to the high amount of incident solar radiation, sunscreens will tend to be used in summer. This will reduce the amount of daylight which enters the room so that artificial lighting costs will be much the same every month of the year. In less sunny London and Copenhagen, however, sunscreens will not usually be required and the need for artificial lighting will drop in summer.

Correct daylighting design will not only reduce energy costs related to artificial lighting but also diminish the possibility of having to use mechanical devices to cool rooms overheated by low-efficiency lighting appliances.

To achieve good daylighting, the glazing must be designed so that there is a correct balance between the heat gains and losses resulting from transmission of thermal radiation in and out of the building and the light entering the building.

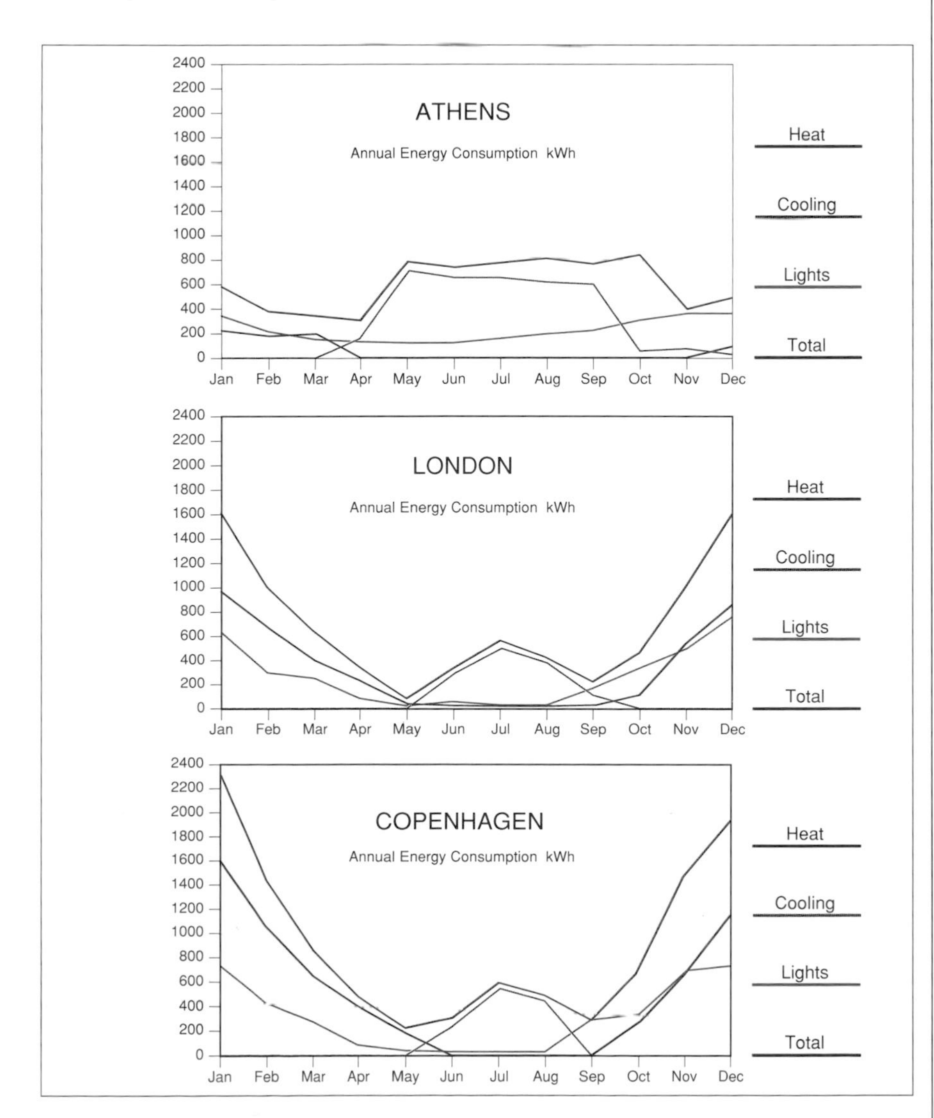

DAYLIGHT AND VISION

VISIBLE LIGHT

What we see as visible light has a wave length in the range 0.38 to 0.76 microns (a micron is a thousandth of a millimetre) - or 380 to 760 nanometres (a nanometre is a millionth of a millimetre). The colour of the 380-420 nanometre band of radiation is deep violet and that of the 660-760 nanometre band is dark red.

Visible light falls into the middle of the total spectrum of electromagnetic radiation. The complete spectrum ranges from cosmic rays (with wave lengths around a millionth of a nanometre) at the lower end to radio waves (with wave lengths around a hundred kilometres) at the upper end. The speed of light is 299,820 kilometres a second.

Daylight is composed of of direct and diffuse light. Direct light beams are, by the time they reach the earth, effectively parallel. Diffuse light is received from the sky after it has been reflected by the gases and water droplets in the atmosphere. For daylighting, the sky can be regarded as the integration of an infinite number of finite point sources.

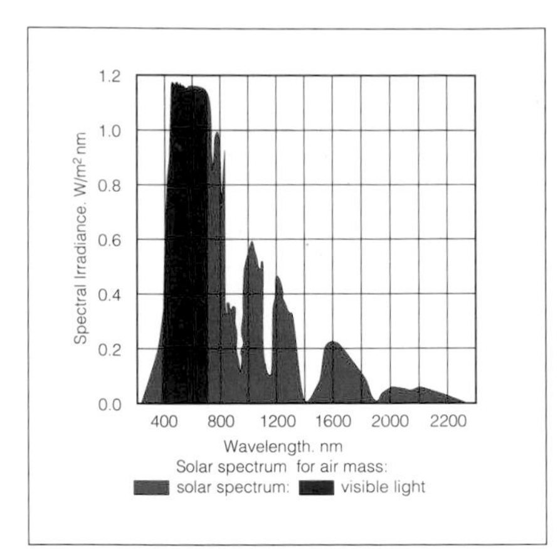

The range of visible light within the solar spectrum.

Color	Wavelenght
Far Violet	380-420 nm
Violet (Purple)	420-440 nm
Blue	440-460 nm
Blue-green	460-510 nm
Green	510-560 nm
Yellow	560-610 nm
Light Red (Orange)	610-660 nm
Dark Red	660-730 nm

Wavebands of light.

VISION AND VISUAL COMFORT

The human eye uses different types of vision to provide the needed sharpness under different levels of illumination. In good light, the eye uses photopic vision, where the cones of the eyes are the principal receptors. At night or when the level of illumination is low, the eye uses scotopic vision, where the rods in the eye are the main receptors.

Colours are identified by photopic vision and during the day the human eye sees best with light with a wavelength of 555 nanometres, which it sees as a yellow similar to the colour of sunlight.

The spectral composition of light influences the way colours are seen by the eye. The reproduction of colours and hence the quality of the light generated by an artificial light source depends on the spectral composition of the source. Different sources create different types of light. Nowadays, artificial light sources are available which offer a spectral composition very close to that of natural light.

For a number of activities, the reproduction of colours in a way which is close to that of daylight (i.e. in a way which can be regarded as neutral) is very important for visual comfort. For other types of activity, this is less significant although it is probably always tiring to have one or two colours predominate.

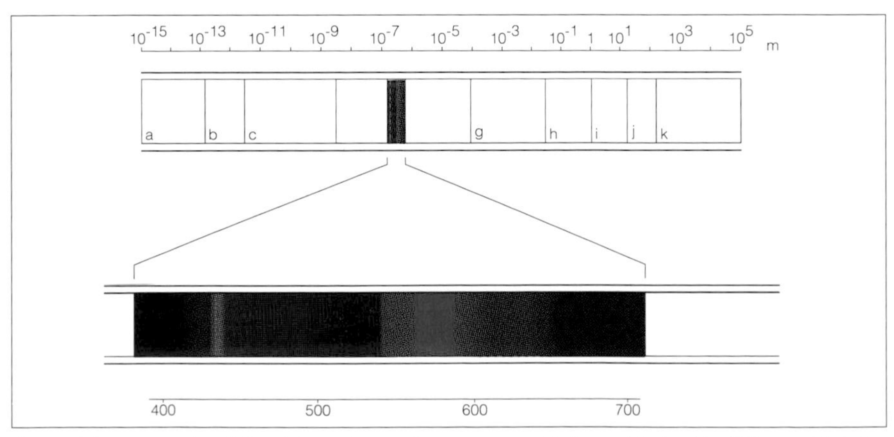

Forms of electromagnetic radiation: (a) cosmic rays; (b) gamma rays; (c) X-rays; (d) ultraviolet (UV) rays; (e) visible light; (f) infrared (IR) waves; (g) radar waves; (h) ultra-high-frequency (UHF) waves; (i) very-high-frequency (VHF) waves.

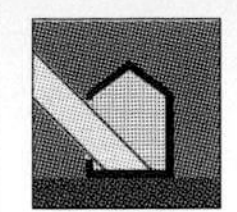

It is important, therefore, to choose light sources for a building which are appropriate to the type of activity to be carried out in each room. For office work, the quality of light needs to be close to that provided by daylight and hence artificial light sources should be selected which provide neutral reproduction of colours.

Visual comfort is also affected by the position of objects in the eyes' field of vision because there are limits to the differences in brightness which can be discerned in different angles of view.
In the centre of the visual field is a solid angle of vision with a 1 degree to 2 degree cone. This is where the most focussed vision (the central foveal vision) occurs.
Within 30 degrees in any direction around the centre line of the foveal vision is the 'near surround'. In this region the eye can readily discriminate between differences in brightness between an object and its background or foreground.
At the extreme edge of binocular vision is the 'far surround', the size and shape of which varies because of the overlap of the visual fields of the right and left eyes.
The ability of the eye to adapt to changes in lighting level and character is very important to the lighting designer. A rule of thumb which is generally applicable is that the eye can easily adjust to the change from bright exterior daylight to an artificially lit room when the artificial lighting level is one hundredth (or more) of the daylight level outside. The figure is determined by the need to adjust not only to the change in brightness level but also to the change in character of the light.
If the move is from bright exterior daylight to a daylight interior, then the eye can cope comfortably with a ratio of 1:200 between the lighting level inside and outside. In either case, once the eye has adapted to the lower lighting level, the ratio can be extended to 1:1000.
It takes about 15 minutes for the eye to adapt to the first 100:1 drop in lighting level. At least 70% of the adjustment is made in the first 90 seconds.

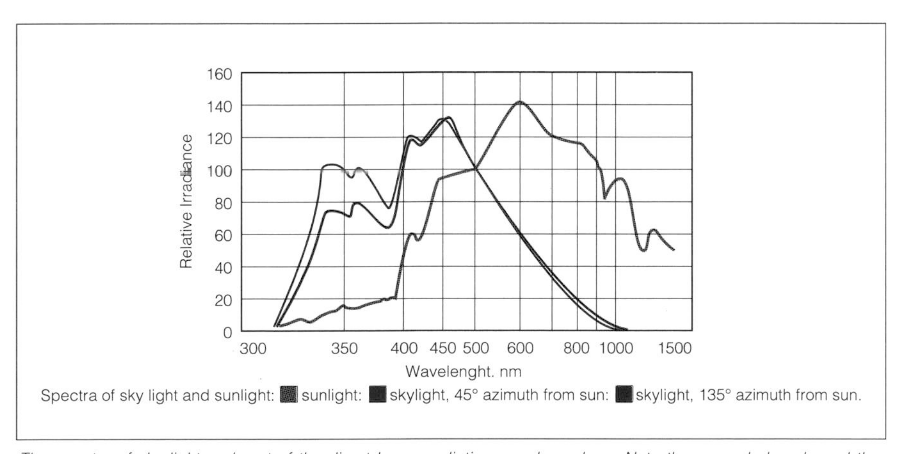

The spectra of sky light and part of the direct beam radiation are shown here. Note the expanded scale and the substantial shift in the shape of the sunlight spectrum in its direct and diffuse components when measured on a clear day.

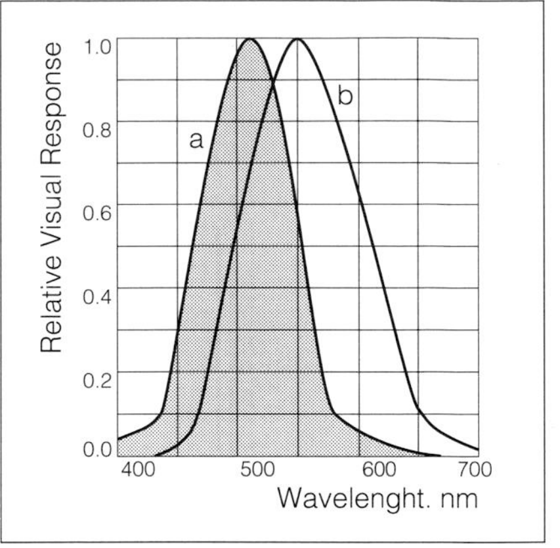

Eye sensitivity curves for (a) scotopic vision, and (b) photopic vision.

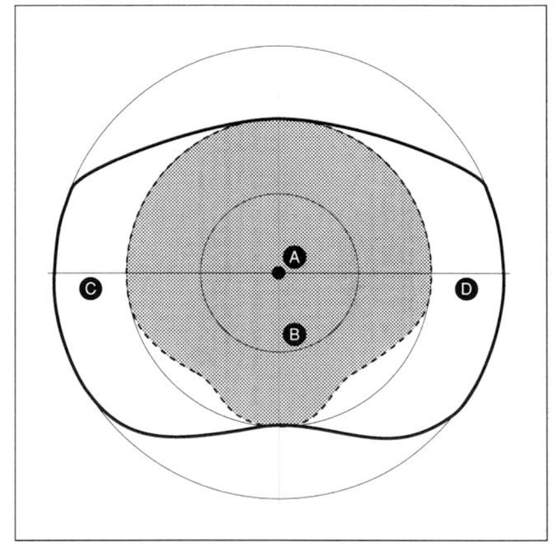

Typical binocular visual field: (a) foveal vision: (b) area seen by both eyes: (c) area seen by left eye: (d) area seen by right eye. Adapted from Kaufman 1974.

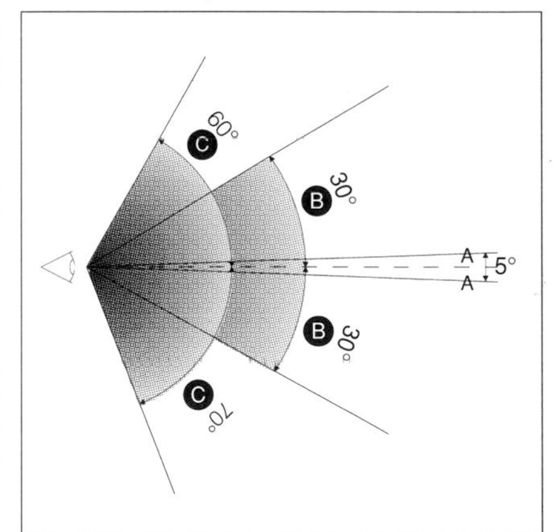

Visual range: (a) foveal vision: (b) near surround: (c) far surround.

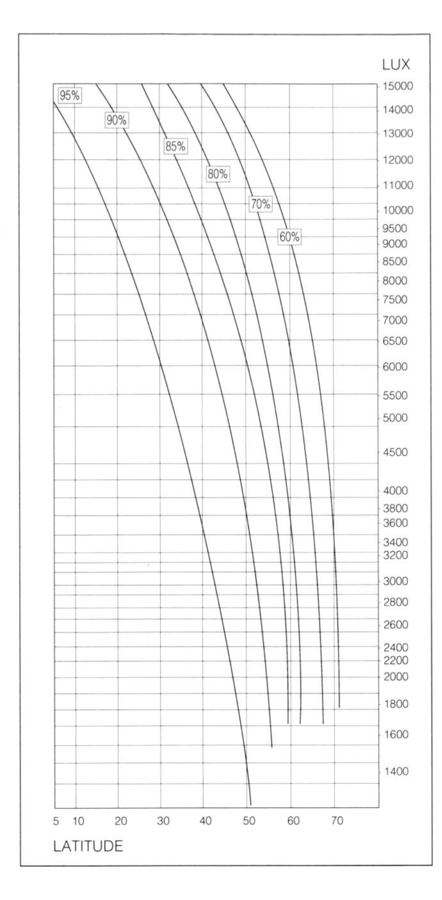

DAYLIGHT AVAILABILITY

As has already been seen in the discussion on solar collection in the heating strategy section of this chapter, the amount of solar radiation incident on a given surface depends on the surface orientation and tilt.

The amount of natural light received in a given outdoor area has three components:

- light coming directly from the sun;
- light received from the sky after being diffused by gases in the atmosphere (this is blue light) and by water droplets in the clouds (white light);
- light from the above two components after it has been reflected by the ground and other nearby surfaces.

Inside a building, a fourth component has to be added: light reflected by interior surfaces.

As a general rule, the term 'daylighting' refers to the light received from the sky excluding direct sunlight. However, under certain circumstances, such as for very sunny climates or for buildings where the contrast produced by direct sunlight makes a significant contribution to the overall quality of the indoor light, it is important to include this direct sunlight component.

The period of time during which daylight is likely to meet the lighting requirements of a building can be calculated for a particular latitude using a set of curves published by the Commission Internationale de l'Eclairage (CIE). From these curves the percentage of the working day during which a required exterior horizontal level of illumination will be reached can be read. The curves do not, however, take into account the individual needs of occupants (some people require higher lighting levels than others), the glare which might be produced by direct sunlight or the effect of shading from external obstacles.

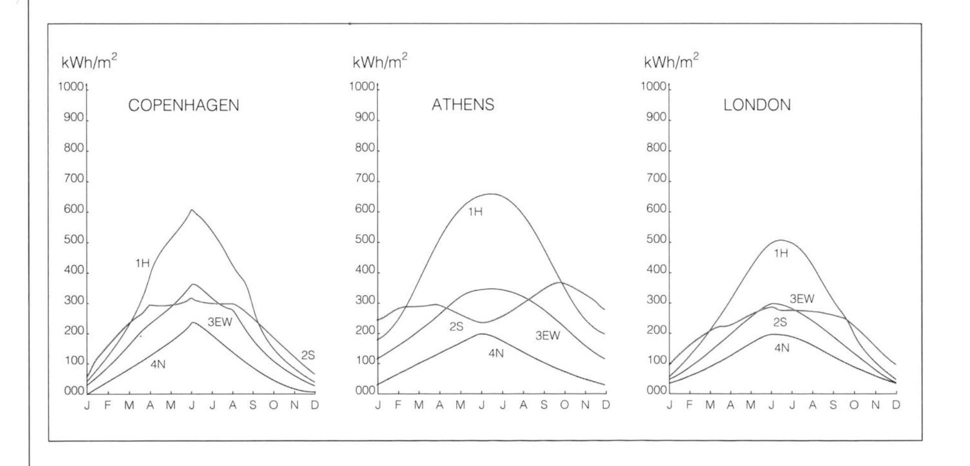

DEFINITIONS AND UNITS

The study of daylighting requires the use of a number of terms, the most common of which are defined here.

LUMINOUS FLUX

Luminous flux is the radiant flux or power emitted by the sun and the sky as viewed by the human eye. Its unit is the lumen. A luminous flux of 680 lumens is produced by a beam of monochromatic radiation of wave length 0.555 microns whose radiant flux is 1 watt. This corresponds to the maximum sensitivity of the human eye.

LUMINOUS EFFICACY OF DAYLIGHT

The luminous efficacy of daylight (i.e. the lumens emitted by a watt of radiant power) is particularly high. It reaches 100 lumens per watt. Artificial light, on the other hand, has a luminous efficacy of only approximately 15 lumens per watt.

ILLUMINANCE

The illuminance at a particular point of a surface is the quantity of luminous flux uniformly distributed over the surface, divided by the area of the surface. The unit of illuminance is the lux, i.e. the illuminance produced on a square metre of surface by a luminous flux of one lumen uniformly distributed over that surface.

LUMINANCE

The luminance of a lit surface is the illuminance received at the surface, modified by the surface's reflectivity. It corresponds to the visual impact on the eye of the surface's luminous intensity and is expressed in candelas per square metre.

EXTERIOR ILLUMINANCE

The exterior illuminance depends on the luminance of the sky, which in turn depends on the sunlight.
The Commission Internationale de l'Eclairage (CIE) has prepared two standard luminance distributions, one for totally overcast conditions and one for clear sky conditions with the direct sunlight excluded. With an overcast sky, the luminance of the sky is independent of orientation. Under clear sky conditions, the luminance depends on the sun's position.
The luminance of the sky varies in monthly and daily cycles. These variations, combined with the meteorological conditions (i.e. the sky clearness), determines the quantity of daylight available for lighting.

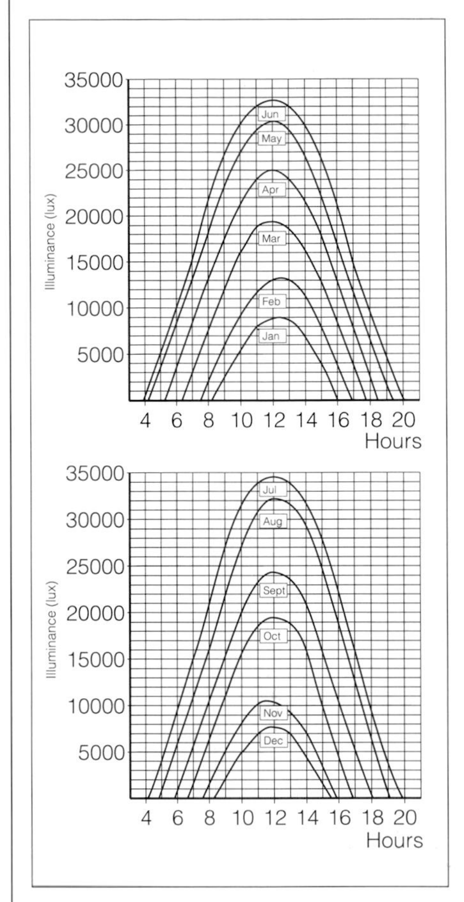

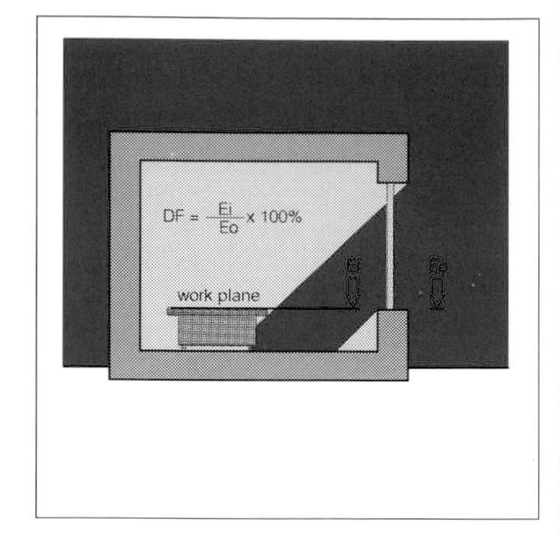

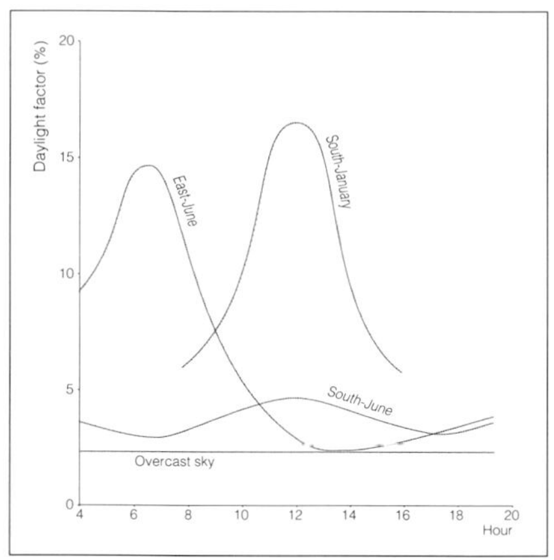

Daylight factor distribution measured at the workplane. Data for Kew (UK).

DAYLIGHT FACTORS

As indicated in the section on daylight availability, the daylight entering a room consists of light coming directly from the sun, light received from the sky and light reflected from the ground and other exterior surfaces. The distribution of the light within the room depends on the size and geometry of the room, the position and characteristics of the glazing and the internal reflections.

Because the exterior illuminance is variable, the daylighting of a room is characterized in terms of a parameter known as the daylight factor (DF). This is the ratio (expressed as a percentage) of the illuminance at a given point on a given plane due to the light received from a sky of known illuminance distribution to the illuminance on the horizontal plane due to an unobstructed hemisphere of this sky. The contribution of direct sunlight to both illuminances is excluded. (The glazing and dirt effects are included.)

The DF has three parts:

- the direct component resulting from the light coming directly from the sky;
- the component resulting from light reflected on outside surfaces;
- the component resulting from reflections within the room.

The DF distribution in a room varies according to the hour of the day, the season of the year, the orientation and the type of sky. Under totally overcast conditions, the DF at a given point is constant because it is independent of orientation, time of day or season of the year.

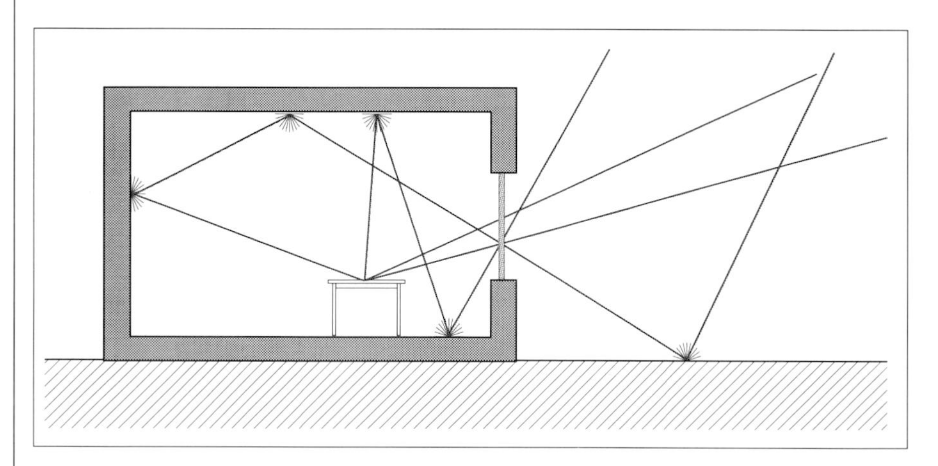

DAYLIGHT DISTRIBUTION

The recommended illuminance level in a room depends on the type of activity taking place there. In office buildings, for instance, the goal is to produce the minimum required illuminance level evenly over the whole of the working surface. Distributing the window openings as uniformly as possible throughout the room can help achieve this. Further information on visual comfort requirements is given in the visual comfort section later in this book.

The penetration and distribution of daylight in a room depends mainly on the size and location of the openings, the type of glazing used, the configuration of the room and the reflections caused by the walls, ceiling and other surfaces.

The internal daylight intensity and daylight factor decrease with distance from the openings and are also affected by the heights of the window sill and head, as shown opposite.

The distribution of daylight in a room is often shown on a room plan by means of isolux and DF curves joining up points of the same illuminance level or DF.

The sketches opposite show typical side and top lighting arrangements for good daylighting distribution.

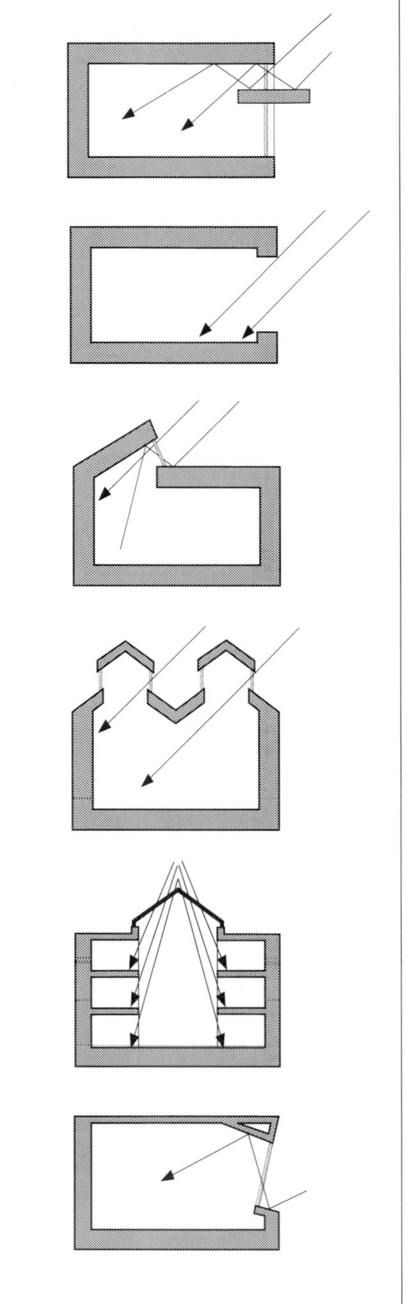

PART 3

THE OCCUPANT

INTRODUCTION

Climate-sensitive architecture cannot exist in isolation from the building occupants. Its purpose is to meet their thermal and visual comfort requirements and the occupants have a part to play in the creation and successful operation of the building. This chapter, therefore, is divided into three parts dealing respectively with thermal comfort, visual comfort and the behaviour of occupants.

The feeling of thermal comfort can be said to be a sense of well-being with respect to heat and it results from a satisfactory interaction of the individual with his or her environment. The section on thermal comfort attempts to evaluate this feeling with respect to the occupant's clothing and activities and the characteristics of his or her surroundings, showing that it is possible to vary the atmosphere in a particular space so that it is appropriate for the activities to be performed there.

In the visual comfort section, attention is paid to achieving the type of lighting in a space which is appropriate to the activities carried out there. Visual comfort needs in a non-domestic building differ from those in dwellings. In offices, for example, the illuminance level is dictated by the requirement that a given task be completed under conditions which optimize the productivity and health of the worker. In housing, the need is to create lighting conditions which are capable of being adapted to the activities to be performed in the different rooms, to the time of day and to individual tastes. The breakdown of a good visual environment into its constituent parts is described together with an indication of how optimum conditions can be achieved for each of them for different types of activity. It is important to consider both the qualitative as well as the quantitative aspects of light.

The final section in this chapter shows how individual occupants can, by their day-to-day behaviour, have a real influence on the successful running of a climate-sensitive building.

CHAPTER 7

THE OCCUPANT
THERMAL COMFORT

The internal temperature of the human body is constant and as the body has no means of storing heat. Any heat generated by it has to be dissipated. An individual's feeling of thermal comfort is optimal when the production of internal heat is equal to the thermal losses from the body. The actual balance between the two depends on seven parameters outlined below.

Activity	Production of metabolic energy M	
	W/m²	met
rest - lying	46	0.8
rest - sitting	58	1.0
office work	70	1.2
house keeping	117	2.0
tennis	290	5.0
squash	406	7.0

Clothing	Clothes thermal resistance C	
	m²K/W	Clo
naked	0.00	0
shorts	0.02	0.1
summer wear	0.08	0.5
winter wear inside	0.16	1
winter wear outside	0.23	1.5

THERMAL COMFORT PARAMETERS

Because of the differences between individuals, it is impossible to specify the precise values of the seven comfort parameters which would give an environment suitable for everyone. The interactions between the parameters have, however, been described by a number of thermal indices (such as the optimal operative temperature, comfort zones, the predicted mean vote and predicted percentage of dissatisfied) which can be used to establish the conditions under which a predicted percentage of occupants will be comfortable - or dissatisfied.

In addition, comfort charts are available to enable a quicker assessment of the comfort zones, for a predicted percentage of the population, (typically 75%) to be made. These show given values of certain comfort parameters as a function of the other comfort parameters.

Clearly, building-related parameters such as thermal inertia and ventilation rate will also have an effect on thermal comfort. Bioclimatic charts exist which show the influence on thermal comfort zones of changing building-related parameters.

The feeling of thermal comfort depends on seven parameters. Three of them - metabolism, clothing and skin temperature - relate to the individual. The other four are linked to the surrounding environment; they are room temperature, relative humidity, surface temperature of the walls and other surfaces in the room and air velocity.

METABOLISM

Metabolism is the sum of the chemical reactions which occur within the body. These reactions are promoted by biological catalysts and the changes resulting from them are usually small. Therefore the amounts of energy required or released at any given stage are minimized. The aim is to maintain the body at a constant internal temperature of 36.7 degrees C. Because the temperature of the body is usually higher than that of the room, the metabolic reactions occur continuously to compensate for loss of heat to the surroundings.

Production of metabolic energy depends on the level of activity in which the individual is engaged. Office work, for instance, generates approximately 0.8 met whereas playing squash produces approximately 7.0 met. The met is the unit of metabolic energy and is equal to 58 watts per square metre. The surface area of the human body, on average, is 1.8 square metres.

CLOTHING

Clothing creates a thermal resistance to the exchange of heat between the surface of the skin and the surrounding atmosphere. The thermal resistance of ordinary summer clothes is 0.5 clo while that of indoor winter wear is 1 clo. The clo is the unit of thermal resistance due to clothes and is equal to 0.155 square metres K per watt.

SKIN TEMPERATURE

The temperature of the skin's surface is a function of metabolism, clothing, room temperature and so on. Unlike internal body temperature, it is not constant.

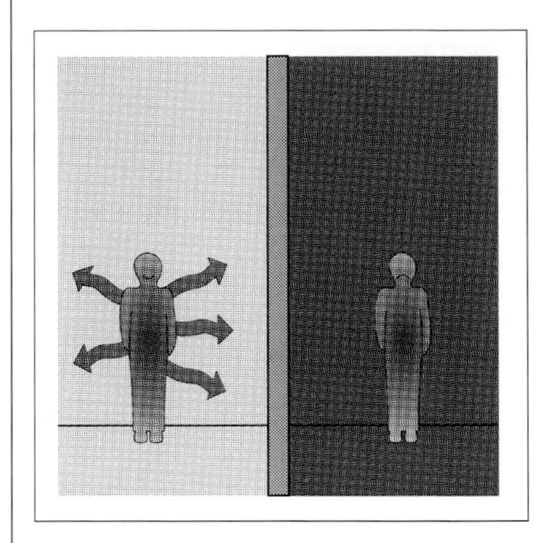

ROOM TEMPERATURE

Room temperature, measured with an ordinary dry bulb thermometer is very important to thermal comfort since more than half the heat lost from the human body is lost by convection to the room air.

RELATIVE HUMIDITY

Relative humidity is the ratio (expressed as a percentage) of the amount of moisture in the air to the moisture it would contain if it were saturated at the same temperature and pressure.

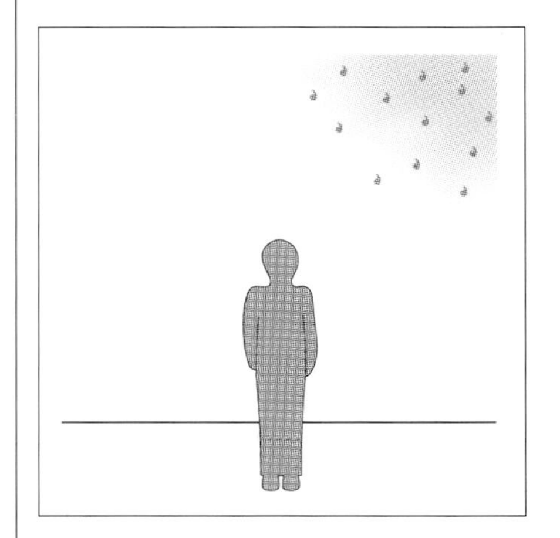

Except for extreme situations (when, for instance, the air is absolutely dry or it is saturated), the influence of relative humidity on the feeling of thermal comfort is relatively small. In temperate regions, for instance, raising the relative humidity from 20% to 60% allows the temperature to be decreased by less than 1 degree C while maintaining the same comfort level. In calculations related to thermal comfort, therefore, the relative humidity is often fixed at 50%.

It is usually recommended that the relative humidity in a room should be somewhere between 20%, to prevent drying up of the mucous membranes, and 80%, to avoid the formation of mould in the building.

SURFACE TEMPERATURE OF PARTITIONS

The average surface temperature of the surfaces enclosing a space is the mean radiant temperature. It influences both the heat lost by radiation from the body to the surfaces and the heat lost by conduction when the individual is in contact with the surfaces.
The radiative heat losses are difficult to quantify because they vary with the individual's position in the room and therefore with the angle between the individual and the surrounding surfaces. As a simplification, the mean radiant temperature can be taken to be the mean of the temperatures of the surrounding surfaces in proportion to their surface areas.

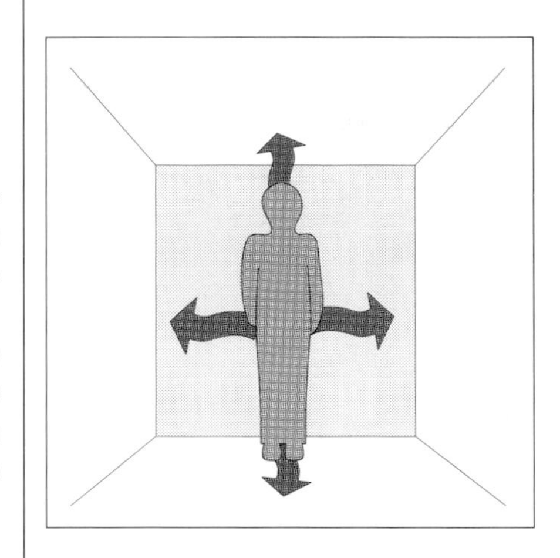

If a building is carefully insulated, the temperature of the internal surface of the outer walls is close to room temperature. This reduces the radiative heat losses and therefore increases the feeling of thermal comfort. It also diminishes the occurrence of convective draughts.

AIR VELOCITY

The velocity of the air relative to the individual influences the heat lost through convection. Within buildings, air speeds are generally less than 0.2 metres per second. The relative air velocity due to the individual's activity can vary from 0-0.1 metre per second for office work to 0.5-2 metres per second for someone playing squash.

Activity	Relative air velocity (V_{air}) m/s
rest, lying	0
rest, sitting	0
office work (sitting)	0.0 - 0.1
house keeping	0.2 - 1.2
tennis	0.5 - 2
squash	0.5 - 2

THERMAL INDICES

As explained above, an individual's feeling of thermal comfort depends on the thermal balance in the room which is, in turn, a function of six parameters - metabolism, clothing, room temperature, relative humidity, surface temperature of the room surfaces and air velocity. (Skin temperature does not come into the equation directly because it is itself mainly a function of metabolism, clothing and room temperature.) A number of thermal indices have been developed for describing the interactions between the six parameters and evaluating the feeling of thermal comfort an occupant is likely to experience in the space.

OPTIMAL OPERATIVE TEMPERATURE

t_{mr} (°C)
air vel ≤ 0.2 m/s
to = 22.5°C
to = 20°C
to = 17.5°C
to = 15°C
to = 12.5°C
to = 10°C
t_{air} (°C)
25 20 15 10 5
10 15 20 25 30

The operative temperature is defined as the uniform temperature of a black radiative enclosure in which the occupant exchanges the same quantity of heat through radiation and convection as he or she would in a non-uniform, real space.
When the air velocity is 0.2 metres per second or less, the operative temperature can be taken to be the mean of the room temperature and the mean radiant temperature.
The optimal value of the operative temperature corresponds to the comfort temperature in the room. Thus, if the comfort temperature has been established to be 20 degrees C, then for a mean radiant temperature of 19 degrees C, the room temperature must be set at 21 degrees C.

COMFORT ZONES

The human body involuntarily regulates its production of internal heat to the thermal conditions of the environment, eventually creating a situation where the metabolic generation of heat is offset by the heat losses so the individual experiences only very small variations in the feeling of thermal comfort and thereby feels at ease.

Thermal feeling	PMV
cold	-3
chilly	-2
fresh	-1
optimal (neutral)	0
(comfort temperature)	
mild	+1
warm	+2
hot	+3

PREDICTED MEAN VOTE

The predicted mean vote (PMV) is a thermal sensation scale. The mean opinion of a large group of individuals expressing a vote on their thermal feeling under different thermal circumstances has been used to provide an index to thermal comfort. A PMV value of zero provides the optimal thermal comfort conditions. A positive PMV value means that the temperature is higher than optimal and a negative value means that it is lower. The comfort zone is generally regarded as stretching from a slight feeling of cold (termed 'fresh', when the PMV is -1) to a slight feeling of warmth (termed 'mild', when the PMV is +1).

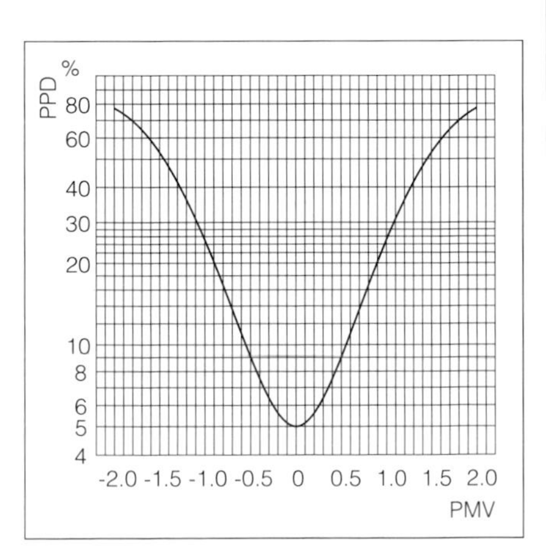

PREDICTED PERCENTAGE OF DISSATISFIED

The predicted percentage of dissatisfied (PPD) is an indication of the percentage of people susceptible to feeling too warm or too cold in a given thermal environment. It can be deduced from the PMV. If, for instance, the PMV is in the range -1 to +1, then the PPD index shows that 25% of the population will be dissatisfied. To reduce this figure to 10%, then the PMV has to be in the range -0.5 to +0.5.

COMFORT CHARTS

A readily accessible way of presenting the relationship between a thermal comfort parameter or thermal index and the other parameters is by means of comfort charts. In one such graphical method, lines show the optimal operative temperature as a function of metabolism (activity level) and clothing for a given PMV and relative humidity.

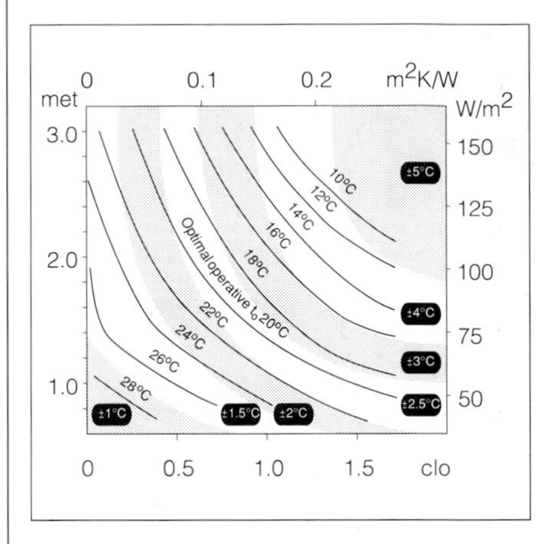

Another approach shows thermal comfort zones in terms of operative temperature and water vapour pressure for a certain percentage of the population for given air velocity, clothing and activity levels.

The diagram opposite gives the optimal operative temperature t_o in terms of the metabolism M and of the clothing C, when taking PMV = 0 and relative humidity (ϕ) = 50%.

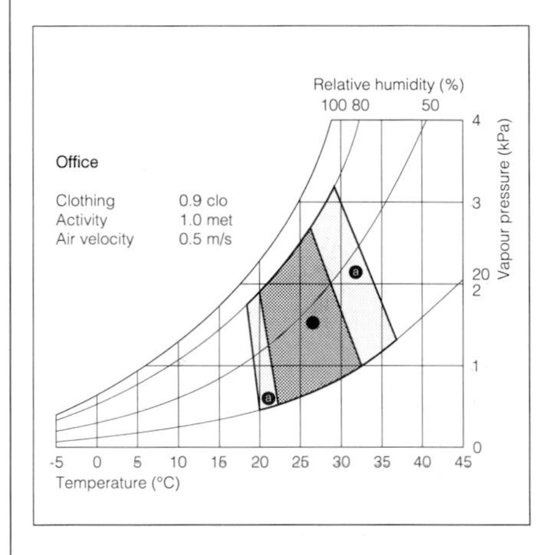

The diagram shows different areas : shaded and non shaded. Each area is characterized by a certain amount of degrees which can be added to or substracted from the optimal temperature, in order to find the limits of the comfort zone corresponding to -0.5 < PMV < +0.5. For example, a person sitting (M = 1.2 met), wearing winter clothes (C = 1 clo) has a comfort temperature of about 21°C. The diagram shows that limits accepted by 90 % of the occupants vary from 18.5°C to 23.5°C, which is 2.5°C around the optimal temperature.

The diagrams of Markus and Morris indicate, for a given value of the metabolism M, of the clothing C and of the relative air velocity V_{air}, the comfort zones for a number of persons. These zones are expressed in terms of the operative temperature and the water vapour pressure.

BIOCLIMATIC CHARTS

Bioclimatic charts have been prepared by Givoni which make it possible to determine the effect on thermal comfort of changing building-related parameters such as thermal inertia and ventilation rate. They show that, by making certain changes to these parameters, the comfort zone can be extended a considerable amount even when the external climate conditions are unfavourable - thus showing that, by applying the concepts of climate-sensitive architecture, the effects of climate changes on the interior environment can be minimized to the extent that they become negligible.

Influence of thermal inertia.

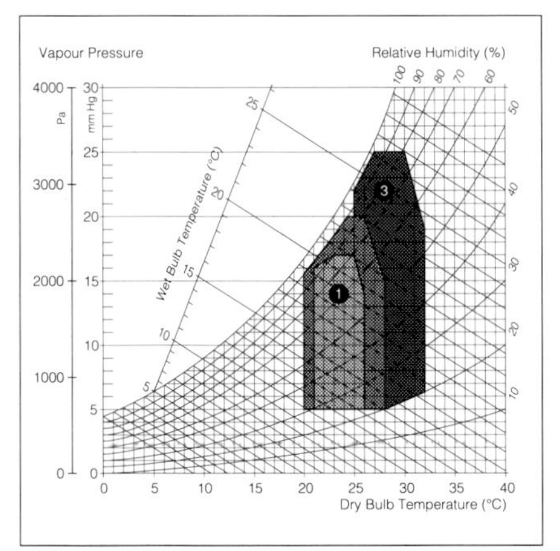

Influence of ventilation.

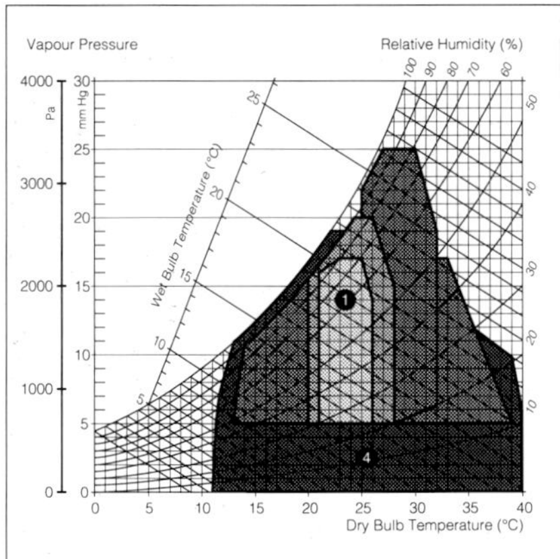

Influence of behaviour.

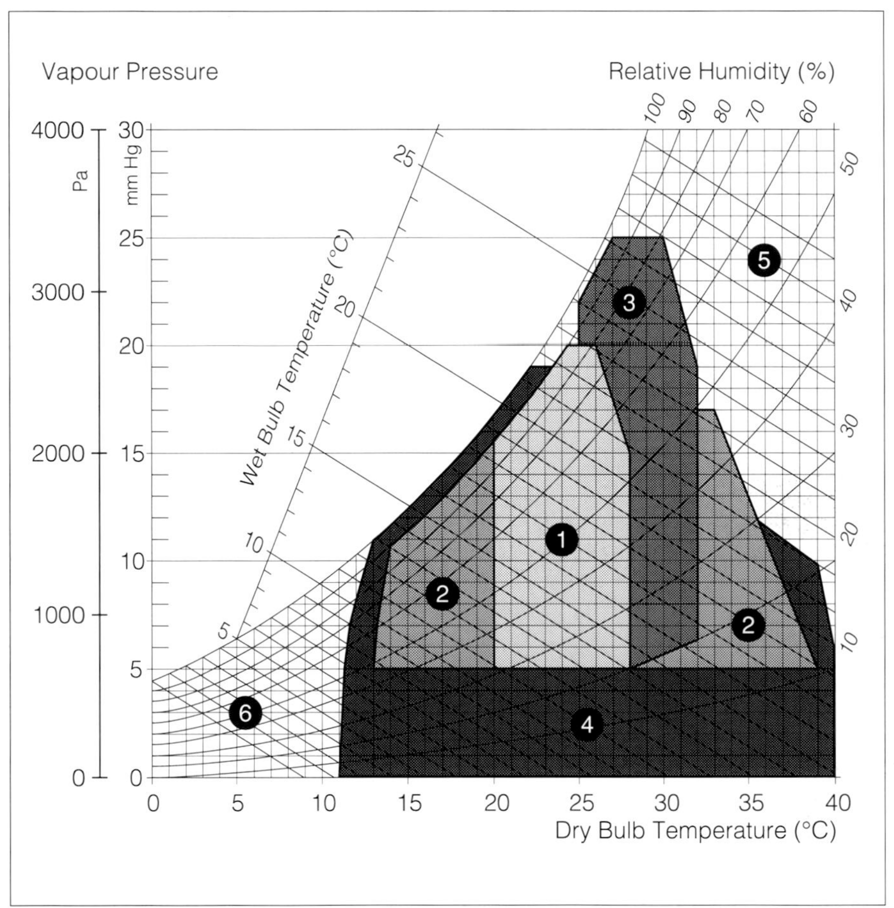

Graph of hygrothermal conditions showing indoor thermal comfort conditions.

1. Comfort zone
2. Zone of influence of thermal inertia
3. Zone of influence of ventilation
4. Zone of influence of occupant behaviour
5. Air conditioning zone
6. Heating zone

CHAPTER 8

THE OCCUPANT
VISUAL COMFORT

Achievement of comfortable lighting conditions in a space depends on the amount, distribution and quality of the light there. Enough illuminance indicated by a sufficiently high daylight factor, should be provided to allow relevant objects to be seen easily, without fatigue.

Recommended illuminance levels

Offices, Workshops and Shops	
storage spaces	150 lux
machine workshops	300 lux
offices	500 lux
drawing offices	750 lux
assembly tasks	1000 lux
precision tasks	1500 lux

Private Houses	
halls	50-100 lux
dining rooms	100 lux
living room, kitchen	200 lux
study	300-500 lux

Recommended daylight factors

Non-Domestic Buildings	
church	1 % minimum
factory	5 % minimum
office	2 % minimum
classroom	2 % minimum
hospital ward	1 % minimum

Private Houses	
bedroom	0.5% at 3/4 of room depth
kitchen	2% at half of room depth
living room	1% at half of room depth

Recommended reflectances

ceilings	0.7-0.85
walls close to light sources	0.6-0.7
other walls	0.4-0.5
floor	0.15-0.3

Luminance ratio

background of visual task: environment	3:1
background of visual task: peripheral field	10:1
light source: adjoining fields	20:1
interior in general	40:1

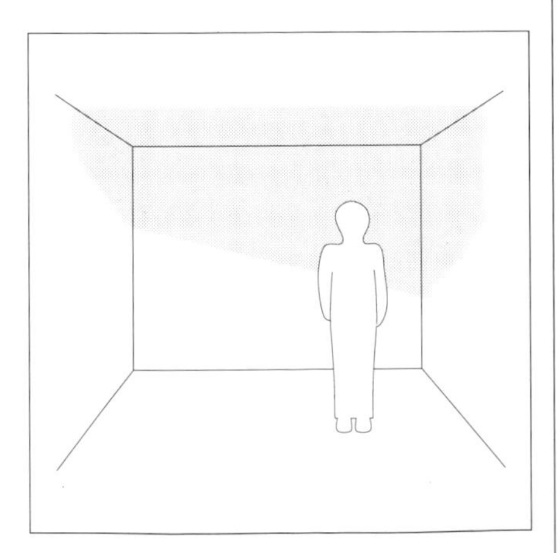

INTRODUCTION

The light distribution in the space should be such that excessive differences in light and shade which could disturb occupants and prevent them from seeing adequately are avoided. Sufficient contrast should, however, be retained for the relief of each object to be brought out. Window openings and artificial light sources should be placed in such a way that glare is minimized

Finally, particular care should be taken over the quality of the light to be provided. Both the spectral composition and light constancy should be appropriate for the task to be performed.

Further information on the spectral composition of light, illuminance and daylight factors can be found in the section on daylight in the building chapter. This present section deals with the level of illuminance required in spaces used for different types of work, the achievement of an appropriate level of contrast, prevention of glare and use of light shelves and other devices to control the light entering a room so that comfortable conditions are provided.

ILLUMINANCE LEVEL

Although the human eye is extremely adaptable, it can nevertheless only perform visual functions within a small range of illuminance levels. For a particular task, the range is affected by the visual performance required, the light distribution in the room and the luminance of the walls and other surfaces.
Recommended optimal illuminance values for the workplane for different types of task, as given in the Building Energy Code published by the (UK) Chartered Institution of Building Services Engineers (CIBSE), are shown opposite.

For daylighting, the illuminance requirements can be translated into minimum values for the daylight factor. These take into account the variability and other properties of daylight. Recommended minimum values for daylight factors for some non-domestic buildings and private dwellings are also shown opposite.

CONTRAST

Contrast is the difference between the visual appearance of an object and that of its immediate background. It can be expressed in terms of luminance, illuminance or reflectivity between surfaces.

The amount and distribution of the light (and hence the amount of contrast) in a room is very dependent on the reflectivity of the walls and other surfaces. It is important, therefore, to choose the wall, floor and ceiling coverings with an eye to their reflectance.

As a general rule, to achieve a correct luminance distribution, light colours should be used for large surfaces and bright ones for smaller surfaces such as furniture, doors, etc.. Recommended reflectances (the ratio of the overall reflected radiant energy to the incident radiant energy) for various inside surfaces are shown opposite.

For comfort, there are limits to the amount of contrast which can be allowed between different parts of a visual field. The recommended maximum values for this, which is also termed the luminance ratio, are also shown opposite.

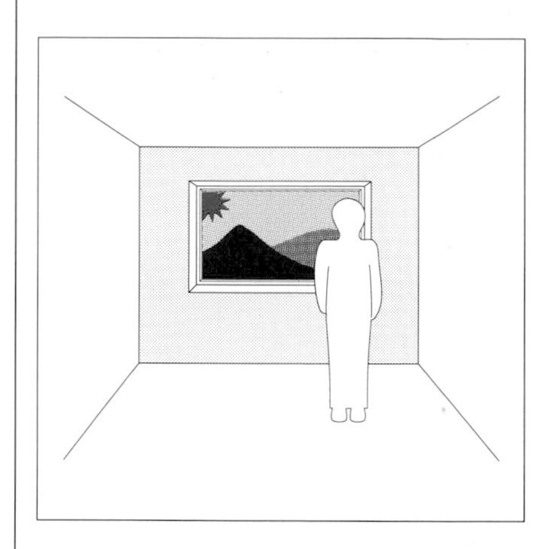

GLARE

Glare is caused by the introduction of a very intense light source into the visual field. It can be mildly distracting or visually blinding for the occupant. Whatever its level, it always produces a feeling of discomfort and fatigue.
Glare can be caused directly, indirectly or by reflection.
Direct glare occurs when a natural or artificial light source with a high luminance enters directly into the individual's field of view. It can be experienced with interior light sources or when the sun or sky is seen through windows either directly or after reflection from an exterior surface.

Indirect glare occurs when the luminance level of walls is too high.
Reflected glare is caused by the specular reflection from light sources on polished interior surfaces.

Glare can be reduced by careful placement of light sources and choosing light sources and backgrounds of suitable luminances.

LIGHT CONTROL

Penetration of solar radiation into a building contributes much to the quality of the lighting there - as long as the sun's rays do not reach the occupants' eyes, either directly or by specular reflection. The penetration of natural light can be controlled in three ways - by reducing the incident flow, the amount of contrast and the luminance of the windows.

Control of direct or diffuse sunlight is important to comfort because it reduces glare. It can be achieved either by incorporation of permanent or movable exterior devices into the building design to reduce the view of the sky or by using movable interior screens to reduce the luminance of the window.

Reduction of excessive contrasts can be achieved by using light-coloured walls and ceilings to give a better light distribution. In particular, light-coloured coverings should normally be used for walls containing window openings.

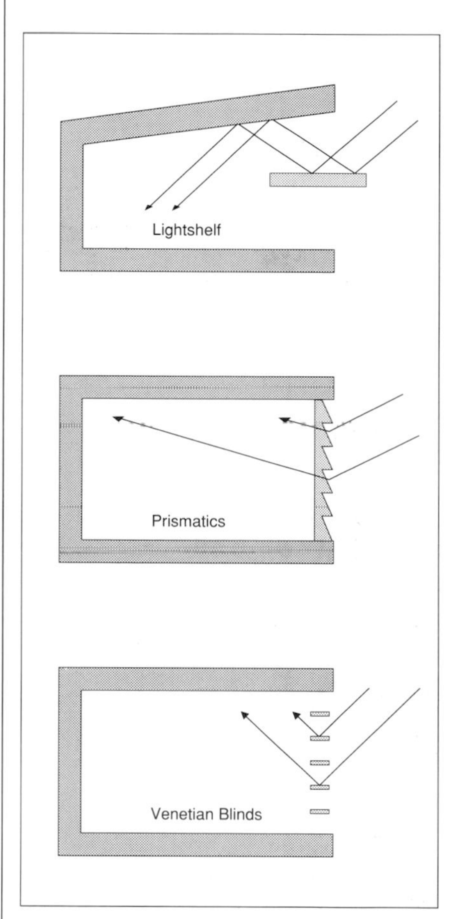

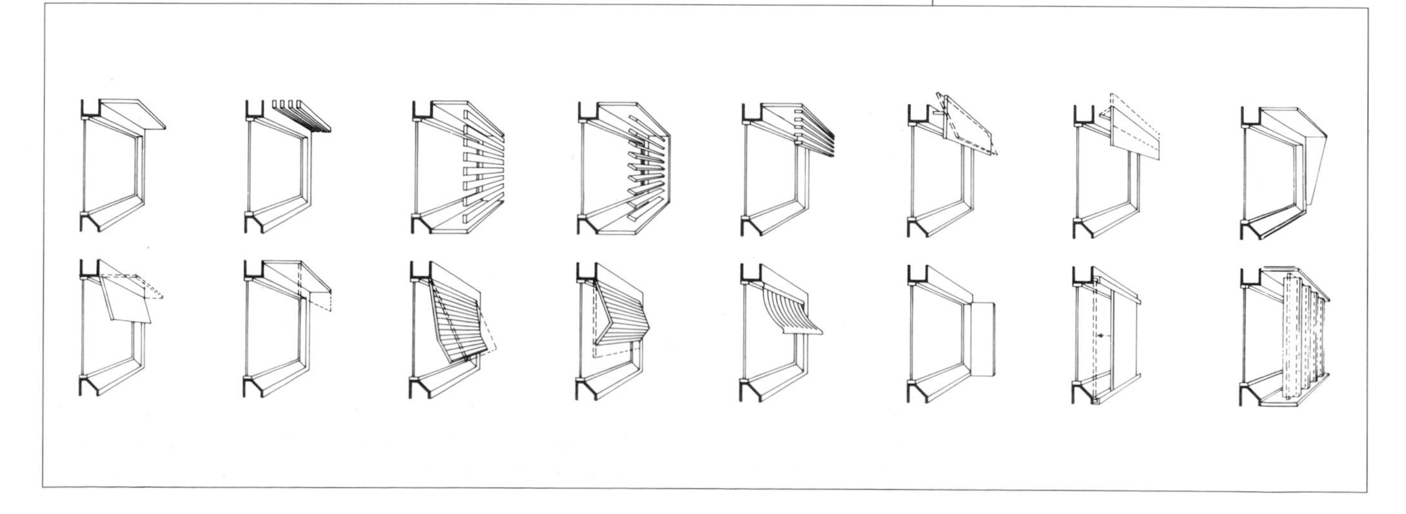

THE OCCUPANT BEHAVIOUR

By their day-to-day behaviour, individual occupants can have a direct influence on the successful running of a climate-sensitive building.

BEHAVIOUR

Few people are as yet fortunate enough to have lived or worked in a climate-sensitive building. Those who have will have found subtle differences between this type of architecture and more 'conventional', highly-serviced buildings. Not only does the climate-sensitive building take less conventional energy to run but it also responds in a more natural way to the needs of occupants and changes in the weather. An understanding of how the building works will enable the occupants to follow with interest the way it responds to climate changes so that they make the best use of the building and the natural and comfortable conditions it has been designed to provide. The designer will realize from the Building chapter what guidelines should be given to the occupant. However, a few examples of how the user can improve the running of the building are given below. Many of the necessary actions can be carried out automatically by straightforward control devices.

In winter, insulating devices such as shutters and curtains can be closed at night to prevent heat loss and opened during the day to allow solar radiation to penetrate the building.

An attached sunspace can be turned into an effective device for collecting solar radiation, when needed, by closing doors and partitions separating this space from the rest of the building. Opening the partitions allows the heat collected to be distributed round the house.
Thermocirculation of the air can be established when required by opening appropriate inlets and outlets.

To prevent overheating, movable protection can be employed and windows opened to create good ventilation.

Care can be taken over the location of appliances (including lighting devices) which create internal gains. Rooms can be organized to make maximum use of daylight. Artificial lighting can be turned off at times when the daylighting is adequate.

If the building spaces are organized into zones with different thermal environments to meet users' needs at different times, the occupants can retire to a small number of well-heated rooms on winter evenings but spread out in milder days to spaces closer to the outside.

Occupants should appreciate the way in which the auxiliary heating and cooling systems run and are regulated. In this way they will be able to make the most of the heating, cooling and daylighting features so that high levels of thermal and visual comfort are maintained throughout the year while energy costs are minimized.

CONCLUSION

This book has attempted to lift the curtain on a somewhat neglected aspect of building design which has considerable potential as a source of architectural inspiration as well as providing important environmental benefits.

The purpose of the book is to inform the designer about general principles and to provide guidance in making professional judgements. There is not always a simple technological solution to a problem - indeed it is a characteristic of most professional activity that decisions are made on the basis of experience, often with incomplete information.

For simplicity and coherence of presentation, the factors concerned with heating, cooling and daylighting design have been outlined consecutively. However, in practice of course these will usually need to be considered together. For example, the sizing of windows to achieve optimal daylighting will have significant implications for the thermal performance of the building.

Thorough analysis of the proposed site and its environment is fundamental to the design of any climate-sensitive building. This analysis will range from the general macroclimatic characteristics of the region, including solar radiation, wind, air temperature and humidity data which are available from meteorological stations, to the more site-specific conditions at the microclimate level which the architect must investigate. This analysis will include the calculation of shadows, the sheltering effects of topography, vegetation or neighbouring buildings and the available daylight.

There is a range of simple physical tools and calculation procedures available for use, and new ones continue to be developed. Computers are often now used to run simple-to-operate but increasingly sophisticated design and evaluation procedures which can take account of interactions between many, sometimes conflicting factors affecting the thermal and daylighting performance of the proposed building.

A second publication entitled "Energy in Architecture: The European Passive Solar Handbook" builds on the principles outlined here and offers more detailed design advice for use at the drawing board.

"We must begin by taking note of the countries and climates in which homes are to be built if our designs for them are to be correct. One type of house seems appropriate for Egypt, another for Spain ... one still different for Rome, and so on with lands and countries of varying characteristics. This is because one part of the Earth is directly under the sun's course, another is far away from it, while another lies midway between these two .... It is obvious that designs for homes ought to conform to diversities of climate."

Vitruvius
First century B.C.

FUNDAMENTALS

SOLAR RADIATION

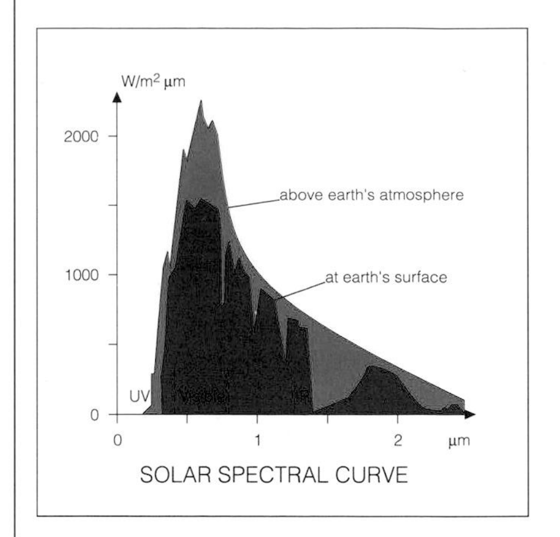

SOLAR SPECTRAL CURVE

The closest fixed star to the earth is the sun. It releases a power flux of 63 million watts per square meter surface area. This energy is produced by nuclear chain reactions. Only part of the emitted energy reaches the outer edge of the earth's atmosphere. This is the ***solar constant***, and it has a value of 1367 W/m^2. The total power received at that level is 220 billion megawatts (2.2×10^{17}W). The solar spectrum extends from about 0.29 µm to 2.5 µm. Less than 2% is contributed to the solar constant from spectral ranges above and below these wavelengths.

The power received at ground level is less because the atmosphere absorbs about 20% of the radiation and reflects about 25%. A fraction of the radiation reaches the ground directly while the remainder reaches it after being diffused by the atmosphere. The global radiation is defined as the sum of direct and diffused radiation.

The amount of energy received depends on the location, the hour of the day, the time of year, and the meteorological conditions.

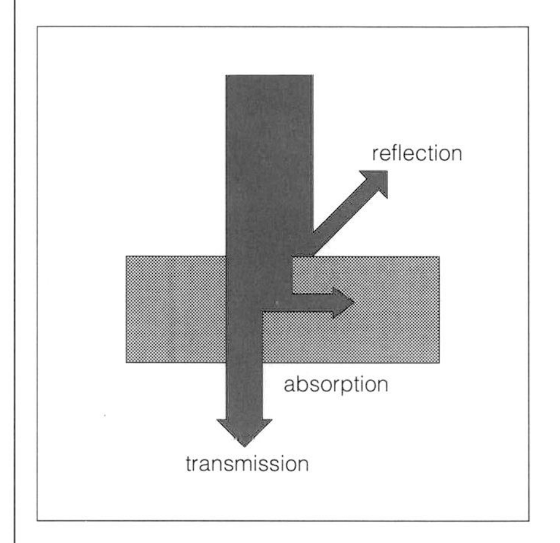

REFLECTION, TRANSMISSION, ABSORPTION

When solar radiation strikes a body:

- part of the radiation is ***reflected***. The fraction reflected depends on the angle at which the sun's rays strike the body and on its surface colour and texture. White and smooth surfaces cause high reflection. Black and rough surfaces cause low reflection. ***Reflectance*** is the ratio of the total reflected flux to the total incident flux.

- another part is ***transmitted***. This happens only if the material is transparent or translucent. ***Transmittance*** is the ratio of the total transmitted flux to the total incident flux.

- the remainder is ***absorbed***. This part is transformed into heat. ***Absorptance*** is the ratio of absorbed flux to the total incident flux.

Materials can be selective with regard to radiation. This means that they do not reflect, absorb or transmit in the same proportions at every wavelength. Thus the materials can also be characterised by their spectral curve of reflectivity, transmissivity or absorptivity, over a range of wavelengths.

HEAT

Heat is one form of energy. It appears as a molecular movement in a body, liquid or gas, or as radiation in space. Heat is measured in Joules as are other forms of energy.

TEMPERATURE

Temperature is a measure of the thermal state of a substance. Different scales exist. The Celsius scale (or centigrade) takes the freezing point of water as 0° and its boiling point as 100° (under standard atmospheric pressure). The Kelvin scale (or absolute temperatures) takes the absolute zero as 0° and keeps the intervals of the Celsius scale. Absolute zero is the lowest theoretically possible temperature. Its value in the Celsius scale is minus 273.15°. Thus x K = x - 273.15°C. A Kelvin degree is an SI unit and is noted K. Temperature changes or differences are more correctly stated in K rather than °C.

SPECIFIC HEAT (C_p)

This is the amount of energy needed to raise a unit mass of a given substance through 1K. It is measured in J/kgK. The specific heat of fluids varies with temperature and pressure.

THERMAL STORAGE CAPACITY

This is the product of the specific heat (C_p) and the density (kg/m^3) of a material and is expressed in J/m^3K.

SENSIBLE HEAT

Sensible heat can be felt or measured. If the sensible heat in an object increases, its temperature also increases, and vice versa, without any change in its state, from solid to liquid for instance.

LATENT HEAT

Latent heat is the heat required to cause a change in the state of a substance from solid to liquid, for instance. This change in state occurs at constant temperature. The same amount of heat is released as is required to reverse the change in state.

THERMODYNAMIC PRINCIPLES

First Principle: Energy Conservation
Energy exists in different forms. It cannot be created or destroyed but only converted from one form to another. Thus in any system the energy input equals the energy output plus the change in stored energy.

Second Principle: Energy Quality and Temperature
Energy transfer takes place spontaneously in one direction only: from a higher level to a lower one. For thermal energy, heat transfer takes place from a warmer body to a cooler one. It is impossible to reverse the direction of heat delivery without any external energy input. The establishment of heat flow requires both a heat source and a heat sink.

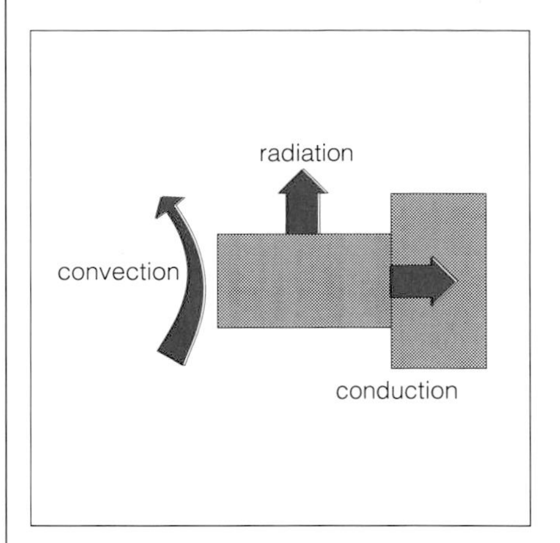

HEAT FLOW

Heat flow is the transfer of thermal energy toward a lower temperature heat sink. This takes place by conduction, by convection or by radiation. The heat flow rate through a body or through space is the amount of energy passing through in unit time, expressed in Joules/sec or Watts. The heat flux density is the heat flow rate per unit area, and is expressed in W/m^2.

CONDUCTION

Heat can be transferred through an object by conduction. Molecular movement is transmitted gradually through an object or between objects in direct contact. The magnitude of heat flow through an object depends on the area of the section perpendicular to the heat flow direction, the thickness of the object, the difference in temperature between the two points being considered, and on the conductivity of the material.

Conductivity (k) is defined as the heat flow rate through a unit area and unit thickness of a material with unit temperature differences across its thickness. The lower the k value, the better its insulating effect. Conductivity is expressed in W/mK.

Resistivity (r) is the reciprocal of conductivity, 1/k = r (mK/W).

Resistance (R) is the product of resistivity and thickness of a body (m^2K/W).

Conductance (U) is the reciprocal of resistance, 1/R = U (W/m^2K). It represents the thermal transmission per square meter of a material, or an assembly of several materials, per degree K of temperature difference between the inside and outside surfaces.

THERMAL INERTIA

Thermal inertia is an expression of the resistance of a body to changes in its temperature. This depends on its thermal storage capacity and its thermal resistance.

THERMAL DIFFUSIVITY

Thermal diffusivity is the rate of heat diffusion throughout a material. The greater the value the greater the heat diffusion. The thermal diffusivity of a material depends on its thermal conductivity, its density and its specific heat. It is measured in m^2/s.

DIURNAL HEAT CAPACITY

The diurnal heat capacity of a material is the daily amount of heat, per unit surface area that is stored and then given back, per unit of temperature change.

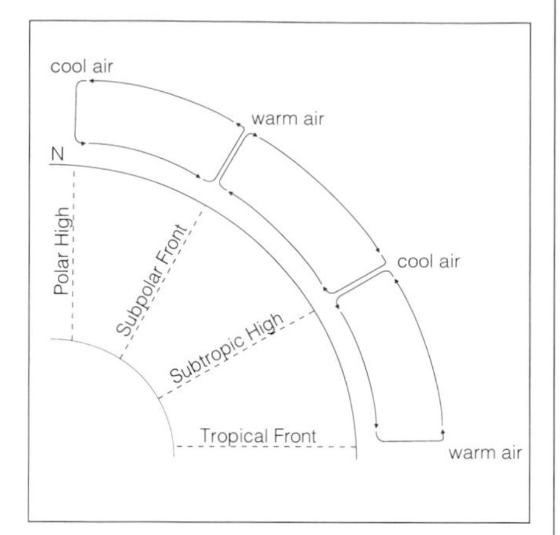

CONVECTION

Convection is heat transfer from the surface of a solid body to a fluid (gas or liquid), or inversely from a fluid to a solid body. The rate of heat flow depends on the area of contact, the temperature difference between solid and fluid, and the convective heat transfer coefficient (or film coefficient), which depends on the flow geometry, the viscosity and velocity of the fluid, and whether the fluid flow is laminar or turbulent.

RADIATION

Heat can be transmitted through space (in a vacuum or a transparent or semi-transparent gas) in the form of radiation from one body to another. The emission wavelength spectrum depends on the nature and on the temperature of the surface of the body. The amount of radiant heat flow depends on the temperatures of the radiating and receiving surfaces and on the emissivity and absorptivity of those surfaces, respectively. Solar energy reaches the earth in the visible band of the radiation spectrum, as well as the longer wavelength infrared and shorter wavelength ultraviolet bands.

Emittance is the ratio of the thermal radiation from a unit area of a surface to the radiation from unit area of a full emitter or "black body" at the same temperature.

PSYCHROMETRIC TERMS

The earth's atmosphere is composed of air (mostly oxygen and nitrogen) and water vapour. The vapour content of air, expressed in grammes of moisture per kg of dry air, is called absolute humidity (g/kg). The maximum quantity of water vapour air can contain depends on the air temperature and is called ***saturation humidity***. It increases with temperature. ***Relative humidity*** is a measure of the quantity of moisture contained in the air, defined as the ratio of the vapour pressure to the saturation pressure at a given temperature.

PSYCHROMETRIC PROCESSES

Dehumidification and condensation by cooling:
For any value of absolute humidity, there is a temperature at which this becomes the saturation humidity. Thus, when cooling air which has a certain moisture, a temperature is reached called ***dew point temperature*** where the amount of vapour is the maximum the air can support, and the moisture begins to condense. Consequently, the quantity of water vapour decreases. This phenomenon can cause condensation problems in non-watertight building elements when the outdoor temperature is lower than the dew point temperature corresponding to the indoor moisture level, causing the moist air leaking out to condense as it cools to the outdoor temperature.

Evaporative Cooling:
The evaporation of liquid water to vapour requires a certain amount of heat. An increase in humidity by evaporation, without any heat input or removal, takes the latent heat necessary from the atmosphere, decreasing its temperature. This process is limited by saturation, so it is most effective in drier climates where more vapour can be added to the air before reaching saturation.

AIR PRESSURE

All fluids exert a pressure due to gravitational force. Contrary to the solid bodies which have only a vertical downward pressure component, the free molecules of fluids transmit the same pressure horizontally. The vertical and horizontal pressure components at one point are equal, and depend on the vertical height of the fluid above and on the fluid density, which in turn depends on temperature.

AIR MOVEMENT

Air masses can be at different temperatures and their pressures can vary from one location to another. By the principle of equilibrium of forces, air from high-pressure areas flows to low-pressure areas tending to equalise the differential pressure and heating of the two air masses. This movement in the atmosphere creates wind.

Solar radiation received at the earth's surface depends mostly on the latitude and the resulting duration of sunshine. Small local variations are induced by cloudiness and atmospheric pollution. At the zone of maximum radiation, air is heated and consequently its density decreases. Warm air masses rise causing a flow toward cooler zones at high level. At high altitude the air becomes cooler and part of this air mass flows down to the surface. The air currents are deflected by the Coriolis forces induced by the earth's rotation. In the northern hemisphere surface air circulation around a high pressure zone (anticyclone) is always clockwise and anticlockwise around a low pressure zone (cyclone). It is the opposite in the southern hemisphere. These broad patterns are modified by local conditions deflecting the wind stream or inducing differences in pressure and temperature.

LUMINOUS FLUX

Luminous flux is the radiant flux or power emitted by the sun and the sky as viewed by the human eye. Its unit is the lumen. A luminous flux of 680 lumens is produced by a beam of monochromatic radiation of wave length 0.555 microns whose radiant flux is 1 Watt. This corresponds to the maximum sensitivity of the human eye.

LUMINOUS EFFICACY OF DAYLIGHT

The luminous efficacy of daylight (i.e. the lumens emitted by a Watt of radiant power) is particularly high. It reaches 100 lumens per Watt. Artificial light, on the other hand, has a luminous efficacy of only approximately 15 lumens per Watt.

ILLUMINANCE

The illuminance at a particular point of a surface is the quantity of luminous flux uniformly distributed over the surface, divided by the area of the surface. The unit of illuminance is the lux, i.e. the illuminance produced on a square metre of surface by a luminous flux of one lumen uniformly distributed over that surface.

LUMINANCE

The luminance of a lit surface is the illuminance received at the surface, modified by the surface's reflectivity. It corresponds to the visual impact on the eye of the surface's luminous intensity and is expressed in candelas per square metre.

EXTERIOR ILLUMINANCE

The exterior illuminance depends on the luminance of the sky, which in turn depends on the sunlight. The Commission Internationale de l'Eclairage (CIE) has prepared two standard luminance distributions, one for totally overcast conditions and one for clear sky conditions with the direct sunlight excluded. With an overcast sky, the luminance of the sky is independent of orientation. Under clear sky conditions, the luminance depends on the sun's position. The luminance of the sky varies in monthly and daily cycles. These variations, combined with the meteorological conditions (i.e. the sky clearness), determines the quantity of daylight available for lighting.

UNITS

Heat capacity	kJ/m^3K (kWh/m^3K)
Air velocity	m/s
Angle	°
Area	m^2
Density	kg/m^3
Energetic exposure	J/cm^2 (Wh/m^2)
Heat flow	W/m^2
Heat quantity	kJ (kWh)
Heat transmission	W/m^2K
Hour	h
Illuminance	lx
Irradiance	W/m^2
Luminance	cd/m^2
Luminous efficacy	lm/W
Luminous flux	lm
Metabolism	W/m^2, met
Solar gain	MJ/m^2 (kWh/m^2)
Temperature	°C, K
Thermal conductivity	W/mK
Thermal resistance	m^2K/W
Volume	m^3

1 W	1 J/sec
1 cal	4,1868 J
1 Wh	$3{\cdot}6.10^3$ Joules
pico (p)	10^{-12}
nano (n)	10^{-9}
micro (μ)	10^{-6}
milli (m)	10^{-3}
kilo (k)	10^3
mega (M)	10^6
giga (G)	10^9
tera (T)	10^{12}

BIBLIOGRAPHY

European Solar Radiation Atlas
Volume I : horizontal surfaces
Volume II : inclined surfaces
Commission of the European Communities
Verlag Tüv Rheinland, 1984

Buildings, Climate and Energy
T.A. Markus and E.N. Morris
Pitman, 1980

Conception thermique de l'habitat
Guide pour la région Provence - Alpes - Côte d'Azur
SOL A.I.R.
Edisud, 1988

Daylighting, Design and Analysis
C.L. Robbins
Van Nostrand Reinhold Company - New York, 1986

Estalvi d'Energia en el dissery d'edificis
Aplicacio de sistemes d'aprofitament solar passiu
Dapartament d'Industria i Energia - Generalitat de Catalunya, 1986

European Passive Solar Handbook - Preliminary edition
Commission of the European Communities
Directorate General XII for Science, Research and Development, 1986

Guide d'aide à la conception bioclimatique
Cellule Architecture et Climat
Services de Programmation de la Politique Scientifique de Belgique, 1986

Man, Climate and Architecture
B. Givoni
Applied Science Publishers - London, 1976

Passive and low energy building design for tropical island climates
ECD Partnership, London, U.K.; Dr N.V. Baker (principal author)
Commonwealth Science Council, 1987

Passive Solar Energy Efficient House Design
Architectural Association School of Architecture; Graduate School,
Energy Studies Programme
Department of Energy Solar Programme - U.K., 1988

Plan Director de la Expo'92
Seminario de Arquitectura bioclimatica
Siciedad Estatal para la exposicion universal de Sevilla 1992, 1987

Thermal comfort
P.O. Fanger
Mc Graw - Hill Book Company - New York, 1973

Working in the City - European Architectural Ideas Competition
S. O'Toole, J O Lewis, Commission of the European Communities
Directorate General XII for Science, Research and Developement.
Gandon, 1988.

INDEX